ESSENTIAL
SOLID MECHANICS

Theory, worked examples
and problems

B.W. Young

Lecturer in Structural
*   and Mechanical Engineering*
University of Sussex

First published 1976 by

THE MACMILLAN PRESS LTD

London and Basingstoke

Associated companies in New York Dublin

Melbourne Johannesburg and Madras

SBN 333 16694 9

To my father

Printed in Great Britain
by Unwin Brothers Limited,
The Gresham Press, Old Woking, Surrey

CONTENTS

PREFACE

The analysis of force, stress and deformation in engineering components is traditionally covered in the study of strength of materials and theory of structures. However, there is so much common ground between these two disciplines, particularly in the early stages, that it seems much more appropriate to treat them as the single subject referred to here as Solid Mechanics.

The purpose of this book is to establish, in concise form, the bases of solid mechanics required by mechanical, civil and structural engineering undergraduates in the first half of a university or polytechnic degree course. The format consists of the elements of the theory for a particular topic followed by a number of worked examples illustrating the application of the theory. Each chapter ends with a selection of problems (with answers) which the student can use for practice. The examples and problems are typical of those set to first and second year undergraduates and have been collected and adapted from a large number of sources during many years of teaching in the field. The precise origin of the questions is unknown but a general acknowledgement is given here. The author alone is responsible for the solutions and answers. There are a total of over two hundred problems in the book of which about half have fully worked solutions.

This book presents all the necessary theory in a compact form and for reasons of space the broader background to the subject has obviously had to be omitted. The student is well advised to extend his knowledge by further reading in the library and a short list of titles recommended for this purpose will be found at the end of the book.

B. W. YOUNG

University of Sussex,
1976

1 FUNDAMENTALS OF EQUILIBRIUM

Before a machine part or a structural member can be put to its required use, the designer has to satisfy himself that it is strong and stiff enough to withstand the loads it is likely to meet during its lifetime.

The purpose of solid mechanics is to describe the way in which applied forces are distributed in a component or structure and to determine the resulting deformations. With this information the designer is able to decide the correct geometry for cross-sections and to select suitable materials.

Forces may be statically or dynamically applied. We shall be concerned almost exclusively with static forces in this book although mention will be made of the effects of suddenly applied loads. The study of dynamic loading in so far as it is associated with time-dependent phenomena requires a separate text of its own.

The techniques and principles we shall be examining are applicable to all materials provided their load-deformation characteristics are known. However, the subject matter of this chapter is quite independent of material behaviour. We shall be concerned here with the force analysis of statically determinate systems. In later chapters the relationships between forces and deformations will be investigated so that the force analysis of statically indeterminate systems and the deformation analysis of both statically determinate and indeterminate systems will be possible. The concept of statical determinacy, which may be unfamiliar, is explained in the following sections.

1.1 THE EQUATIONS OF STATICAL EQUILIBRIUM

We shall make a start by looking at the requirements for equilibrium of a body subjected to a general load system consisting of forces and moments.

The concept of force is readily understood. We recognise that a force has magnitude, direction and a definite line of action. The concept of moment is less obvious. The moment of a force about a point is defined simply as the product of the force and the perpendicular distance of its line of action from that point.

An unrestrained body that is acted upon by a system of forces and moments will move. In the three spatial dimensions this movement will consist in general of three components of translation along three axes mutually at right angles and three components of rotation about these axes. The body therefore has six degrees of freedom of movement. If we confine our attention to two dimensions, the situation is simpler since the body now has three degrees of freedom: two components of translation along two axes at right angles and one rotation in the plane.

An obvious requirement for a satisfactory structure is that it should be supported in such a way that movement as a whole under the action of applied forces is prevented.

The three basic types of structural support are shown in figure 1.1. The roller support in figure 1.1a allows the body to rotate about a pin or hinge (shown as a small circle) and to move horizontally. Vertical movement is prevented by the reaction R_v. The pinned support in figure 1.1b allows the body to rotate but the horizontal and vertical components of translation are prevented by the two independent reactions R_v and R_h. The three independent reactions (two forces, R_h and R_v and the moment M) provided by the built-in support shown in figure 1.1c, prevent rotation of the body and translation in the vertical and horizontal directions. To restrain a body completely in two dimensions we therefore require a minimum of three independent reactions which may consist of two forces at right angles and a moment, or three forces that combine to give the same effect as two forces and a moment.

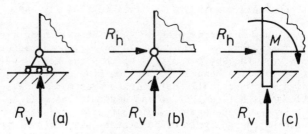

Figure 1.1

The body is said to be in statical equilibrium if the applied forces and the reactions balance each other. This situation is satisfied for a two-dimensional body if

(i)   the sum of the forces in the x-direction is zero
(ii) the sum of the forces in the y-direction (at right angles to the x-direction) is zero, and
(iii) the sum of all the moments about any point in the xy-plane is zero.

These three conditions give rise to three equations of statical equilibrium.

If a body is supported in such a way that three independent reactions are required, it is said to be statically determinate with respect to the supports since the three equations of statical equilibrium are sufficient to determine the magnitudes of the reactions. If more than three independent reactions are present their values cannot be found solely from the three equations and the body is then said to be statically indeterminate with respect to the supports.

The following example illustrates the application of the equilib-

2

rium equations to the determination of reactive forces in two dimensions for a statically determinate body.

Example 1.1

Figure 1.2 shows a box-girder bridge section being gradually raised
into position using two lifting-cables attached at corners A and B
and a tethering cable attached at corner C. The purpose of the
tethering cable is to maintain the box in a horizontal position.
The box is 10 m long, 2 m deep and the centre of mass is at 7 m from
the left-hand end. The weight of the box is 200 kN. Determine the
cable tensions when the cables attached at A, B and C make angles
with the horizontal of 60°, 30° and 45° respectively.

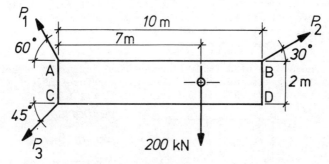

Figure 1.2

  (i) Summing horizontal forces

$$P_1 \cos 60° + P_3 \cos 45° - P_2 \cos 30° = 0$$

therefore

$$P_1 + \sqrt{2}P_3 - \sqrt{3}P_2 = 0 \tag{1}$$

  (ii) Summing vertical forces

$$P_1 \sin 60° + P_2 \sin 30° - P_3 \sin 45° - 200 = 0$$

therefore

$$\sqrt{3}P_1 + P_2 - \sqrt{2}P_3 - 400 = 0 \tag{2}$$

  (iii) Summing moments about A

$$(200)(7) - 10P_2 \sin 30° + 2P_3 \cos 45° = 0$$

therefore

$$1400 - 5P_2 + \sqrt{2}P_3 = 0 \tag{3}$$

Eliminating P_1 from equations 1 and 2 we have

$$400 - 4P_2 + \sqrt{2}(1 + \sqrt{3})P_3 = 0 \tag{4}$$

Solving equations 3 and 4 simultaneously gives

3

$P_2 = 354 \cdot 5$ kN and

$P_3 = 263 \cdot 5$ kN

and by substitution in equation 1 we have

$P_1 = 241 \cdot 4$ kN

1.2 STATICAL DETERMINACY OF PIN-JOINTED FRAMES

We have been referring to the forces and reactions on a body. If
the body is a structure or part of a structure it will usually be
composed of an assemblage of members and it is necessary to check
the degree of determinacy for the whole structure. To explore the
idea of over-all statical determinacy we shall examine the two-
dimensional, or plane, pin-jointed frame. Subsequently we shall
look at pin-jointed space frames.

A plane pin-jointed frame is composed of straight members carrying axial forces alone. Each joint is therefore subject to two
components of translation in directions at right angles. The first

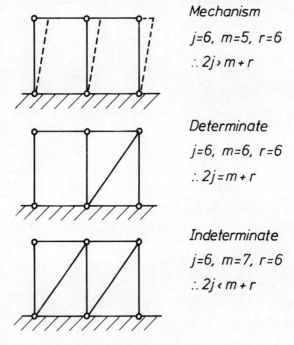

Mechanism

$j=6$, $m=5$, $r=6$

$\therefore 2j > m + r$

Determinate

$j=6$, $m=6$, $r=6$

$\therefore 2j = m + r$

Indeterminate

$j=6$, $m=7$, $r=6$

$\therefore 2j < m + r$

Figure 1.3

two conditions of statical equilibrium may therefore be applied at
each joint. If there are j joints in the frame this will provide $2j$
equations of equilibrium. In each member there is an unknown axial
force and there are also a number of unknown independent reactions.
If the number of members in the frame is m and the number of inde-

4

pendent reactions is r, there are a total of $m + r$ unknown forces to be calculated. The frame is statically determinate if the number of equations of equilibrium are equal to the number of unknown forces and reactions, thus

$$2j = m + r \qquad\qquad (1.1)$$

This equation must be satisfied by the whole frame or any portion disconnected from the whole frame.

Figure 1.3 illustrates the use of equation 1.1 to evaluate the degree of determinacy for a plane pin-jointed frame.

Example 1.2

Determine the reactions at the wall for the pin-jointed cantilever frame shown in figure 1.4.

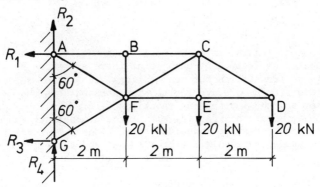

Figure 1.4

There are seven joints, ten members and four independent reactions. Equation 1.1 is therefore satisfied and the frame is statically determinate.

(i) Summing the horizontal forces we have

$$R_1 + R_3 = 0 \qquad\qquad (1)$$

(ii) Summing the vertical forces we have

$$R_2 + R_4 = 60 \text{ kN} \qquad\qquad (2)$$

(iii) Summing the moments about A we have

$$R_3 \,(4/\sqrt{3}) + 20(2) + 20(4) + 20(6) = 0 \qquad\qquad (3)$$

From equation 3

$$R_3 = -\ 60\sqrt{3} \text{ kN}$$

and from equation 1

$$R_1 = + 60\sqrt{3} \text{ kN}$$

The negative sign for R_3 shows that the force is acting in the opposite direction to that initially assumed. It now appears that we have no way of determining the separate values of R_2 and R_4. However, there is the additional piece of information that the line of action of the resultant of R_3 and R_4 must be directed along GF since the force in GF is axial, thus

$$R_4 = - R_3 \tan 30° \tag{4}$$

hence
$$R_4 = 60 \text{ kN}$$

and $R_2 = 0$

It follows that the force at G provided by the member GF is directed at the joint and has the value given by

$$F_{GF} = \sqrt{(R_3^2 + R_4^2)} = 120 \text{ kN}$$

The argument leading to equation 1.1 can easily be extended to a pin-jointed space frame. At each joint there are three possible components of translation but no rotations, thus three force-equations of equilibrium are available. The unknowns are the forces in each member and the independent reactions offered by the supports. Remember that we are now dealing with three-dimensional supports so that a pin provides three independent reactions (all forces) and a built-in support provides six (three forces and three moments). The three-dimensional equivalent of equation 1.1 is thus

$$3j = m + r \tag{1.2}$$

1.3   FORCE ANALYSIS OF PIN-JOINTED PLANE FRAMES

As we have seen, each joint in a plane frame is subjected to a system of concurrent member forces. These forces may be evaluated by systematic resolution at each joint in turn provided the frame is statically determinate. The resolution may be performed graphically by drawing force polygons or analytically by solving sets of simultaneous equations. Graphical methods will not be considered here; instead we shall examine three alternative analytical methods.

1.3.1   Joint Resolution

Consider the simple frame shown in figure 1.5. Since only vertical forces are applied to the frame, the horizontal reaction, R_3, is zero. By resolving vertically and taking moments about C we have

$$R_1 + R_2 = 32 \text{ kN}$$

$$4R_2 = 72 \text{ kN m}$$

Hence

6

$R_1 = 14$ kN, $R_2 = 18$ kN.

We shall assume for convenience that all members are in tension (being stretched) so that as far as a joint is concerned, all member forces are directed away from it (see figure 1.5).

Although any joint may be chosen to start with it is simpler to select a joint with not more than two unknown forces such as joints

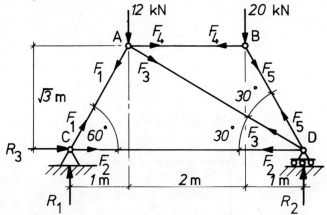

Figure 1.5

B and C in figure 1.5. Supposing we start with joint C, resolving horizontally

$$F_1 \cos 60° + F_2 = 0 \qquad\qquad (a)$$

Resolving vertically

$$F_1 \sin 60° + 14 = 0 \qquad\qquad (b)$$

Hence

$$F_1 = -28/\sqrt{3} = -16 \cdot 2 \text{ kN}$$

$$F_2 = +14/\sqrt{3} = +8 \cdot 1 \text{ kN}$$

The negative sign indicates that member AC is in compression (being squashed).

Moving now to joint B we have

$$F_4 - F_5 \cos 60° = 0 \qquad\qquad (c)$$

$$F_5 \cos 30° + 20 = 0 \qquad\qquad (d)$$

Hence

$$F_5 = -40/\sqrt{3} = -23 \cdot 1 \text{ kN}$$

7

$$F_4 = -20/\sqrt{3} = 11 \cdot 5 \text{ kN}$$

Finally at joint A by resolving horizontally we have

$$F_4 + F_3 \cos 30° - F_1 \cos 60° = 0 \tag{e}$$

Hence $F_3 = + 4$ kN.

1.3.2 The Method of Sections

If the whole frame is in equilibrium, sections cut from the frame will remain in equilibrium provided the member forces are acting on the cut ends.

Suppose members AB, AD and CD are cut and the right-hand side of the frame is removed. We should then have the situation shown in figure 1.6.

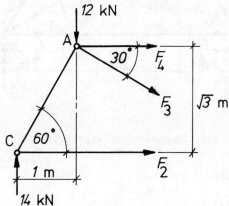

Figure 1.6

The forces placed on the cut ends must maintain the equilibrium of the part frame under its applied loads, thus resolving horizontally

$$F_4 + F_2 + F_3 \cos 30° = 0 \tag{a}$$

Resolving vertically

$$12 + F_3 \cos 60° - 14 = 0 \tag{b}$$

Taking moments about A

$$(14)(1) - \sqrt{3} \, F_2 = 0 \tag{c}$$

From equation c, $F_2 = 14/\sqrt{3} = 8 \cdot 1$ kN

From equation b, $F_3 = + 4 \cdot 0$ kN

From equation a, $F_4 = -20/\sqrt{3} = -11 \cdot 5$ kN

8

These results are, of course, the same as those found by the previous method. One further cut would be necessary to determine the remaining forces, F_1 and F_5.

The method of sections is useful if particular member forces are required. The sometimes tedious process of working from joint to joint is then avoided.

1.3.3 Tension Coefficients

Consider the equilibrium of a typical plane-frame joint such as that shown in figure 1.7.

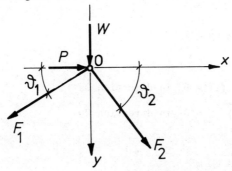

Figure 1.7

P and W are applied forces at the joint, F_1 and F_2 are member forces and Ox and Oy are horizontal and vertical axes respectively. Resolving horizontally and vertically we have

$$P + F_2 \cos \theta_2 - F_1 \cos \theta_1 = 0$$

and $W + F_1 \sin \theta_1 + F_2 \sin \theta_2 = 0$

Now let $F_1 = t_1 L_1$ and $F_2 = t_2 L_2$ where t_1 and t_2 are tension coefficients and L_1 and L_2 are the member lengths. The equilibrium equations become

$$P + t_2 (L_2 \cos \theta_2) - t_1 (L_1 \cos \theta_1) = 0$$

and $W + t_1 (L_1 \sin \theta_1) + t_2 (L_2 \sin \theta_2) = 0$

But $L_1 \cos \theta_1 = x_1$, the projection of L_1 on Ox

$L_2 \cos \theta_2 = x_2$, the projection of L_2 on Ox

$L_1 \sin \theta_1 = y_1$, the projection of L_1 on Oy

$L_2 \sin \theta_2 = y_2$, the projection of L_2 on Oy

The final form of the equilibrium equations is thus

$$P + t_2 x_2 - t_1 x_1 = 0 \quad \text{in direction } Ox$$

9

and $W + t_1y_1 + t_2y_2 = 0$ in direction Oy

We shall now apply the method of tension coefficients to the frame shown in figure 1.5. The origin of the x- and y-axes is placed at the joint under consideration. At joint C

$$t_1 (1) + t_2 (4) = 0 \quad \text{in direction } Ox$$

$$t_1 (\sqrt{3}) + 14 \quad = 0 \quad \text{in direction } Oy$$

therefore

$$t_1 = - 14/\sqrt{3} \text{ kN m}^{-1}$$

and $t_2 = + 7/2\sqrt{3}$ kN m^{-1}

At joint B

$$t_4 (2) - t_5 (1) = 0 \quad \text{in direction } Ox$$

$$20 + t_5 (\sqrt{3}) \quad = 0 \quad \text{in direction } Oy$$

therefore

$$t_5 = - 20/\sqrt{3} \text{ kN m}^{-1}$$

and $t_4 = - 10/\sqrt{3}$ kN m^{-1}

At joint A
$$t_1 (1) - t_4 (2) - t_3 (3) = 0 \quad \text{in direction } Ox$$

therefore

$$t_3 = + 2/\sqrt{3} \text{ kN m}^{-1}$$

The calculation of the member forces from the tension coefficients is best shown in tabular form as follows.

Member	t (kN m^{-1})	L (m)	$F = tL$ (kN)
AC, (1)	$- 14/\sqrt{3}$	2	$- 28/\sqrt{3}$
CD, (2)	$+ 7/2\sqrt{3}$	4	$+ 14/\sqrt{3}$
AD, (3)	$+ 2/\sqrt{3}$	$2\sqrt{3}$	$+ 4$
AB, (4)	$- 10/\sqrt{3}$	2	$- 20/\sqrt{3}$
BD, (5)	$- 20/\sqrt{3}$	2	$- 40/\sqrt{3}$

The negative sign indicates compression. Members in compression are called struts and those in tension are called ties.

The use of tension coefficients avoids the need to calculate angles and their sines and cosines. The advantage over the method of joint resolution (discussed in section 1.3.1) is marginal in the case of plane frames but becomes significant for space frames.

1.4 FORCE ANALYSIS OF PIN-JOINTED SPACE FRAMES

The use of tension coefficients is a powerful method of dealing with the force analysis of space frames. The principles have been intro- duced in section 1.3.3 where they were applied to a plane frame, but the method can be extended without difficulty into three dimensions. This is best illustrated by a worked example.

Example 1.3

Determine the forces in the pin-jointed tripod cantilever bracket shown in figure 1.8.

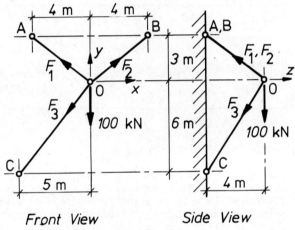

Figure 1.8

There are four joints, three members and nine independent react- ions, thus equation 1.2 is satisfied and the tripod is statically determinate.

The three member forces may be determined by examining the equi- librium of the joint O. In terms of tension coefficients the force equilibrium equations are, in direction Ox

$$4t_1 + 5t_3 - 4t_2 = 0 \tag{1}$$

in direction Oy

$$3(t_1 + t_2) - 6t_3 - 100 = 0 \tag{2}$$

in direction Oz

$$4(t_1 + t_2 + t_3) = 0 \tag{3}$$

11

These equations may be solved to give

$$t_1 = + 25/2 \text{ kN m}^{-1}$$

$$t_2 = - 25/18 \text{ kN m}^{-1}$$

$$t_3 = - 100/9 \text{ kN m}^{-1}$$

The member lengths are determined from their projections as follows

$$L_1 = L_2 = \sqrt{(3^2 + 4^2 + 4^2)} = \sqrt{41} \text{ m}$$

$$L_3 = \sqrt{(5^2 + 6^2 + 4^2)} = \sqrt{77} \text{ m}$$

Using a tabular presentation for the force calculation we have

Member	t (kN m^{-1})	L (m)	$F = tL$ (kN)
OA, (1)	+ 25/2	$\sqrt{41}$	+ 25$\sqrt{41}$/2 = +80
OB, (2)	- 25/18	$\sqrt{41}$	- 25$\sqrt{41}$/18 = -8·9
OC, (3)	- 100/9	$\sqrt{77}$	- 100$\sqrt{77}$/9 = -97·5

1.5 SHEAR FORCE AND BENDING MOMENT

In general, at a particular point in a structural member subjected to a two-dimensional load system there will be a moment, an axial force, and a shear force which acts at right angles to the axial force. The shear force is so called because it tends to shear, or cut, the member, just like the action of a pair of scissors on a strip of paper or cloth. The effect of the moment is to cause bending of the member and it is therefore referred to as a bending moment.

Similarly, a member subjected to a three-dimensional load system will develop at a point two shear forces, an axial force, two bending moments and a torque.

We shall confine our attention here to straight members without axial forces (usually called beams) subjected to a two-dimensional load system. Thus we shall be considering a single shear force and a single bending moment to be acting at a particular point.

The following definitions of shear force and bending moment are consistent with a statement of the equilibrium of the beam.

(1) The shear force at a point in a beam is the algebraic sum of all the forces to one side of the point which act at right angles to the beam axis.

(2) The bending moment at a point in a beam is the algebraic sum

12

of all the moments of forces and concentrated moments to one side of the point.

As an example we shall determine the shear force and bending moment at point B in the beam shown in figure 1.9.

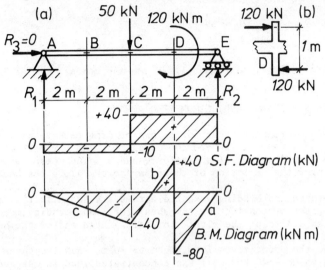

Figure 1.9

The loading on the beam consists of a concentrated force of 50 kN at C and a concentrated clockwise moment or couple at D. Such a moment might be applied by a pair of equal and opposite horizontal forces acting on each side of D (see figure 1.9b).

To determine the unknown reactions R_1 and R_2, we resolve vertically and take moments about A, thus

$$R_1 + R_2 = 50 \text{ kN}$$

and $8R_2 = 4 (50) + 120 = 320 \text{ kN m}$

therefore

$$R_1 = 10 \text{ kN}$$

and $R_2 = 40 \text{ kN}$

The convention adopted for shear force is that to the right of the point, upward forces are positive and to the left, downward forces are positive. Therefore, determining the shear force (Q_B) at B by summing vertical forces to the right, we have

$$Q_B = -50 + R_2 = -10 \text{ kN}$$

If we sum the forces to the left of B we obtain the same result.

The convention adopted for bending moment is that positive moments

13

cause the beam to bend convex upwards. Such a moment is referred to as 'hogging'. The negative bending—moment causing the beam to bend convex downwards is referred to as a 'sagging' moment. If we determine the bending moment (M_B) at point B in figure 1.9 by summing all the moments to the right we have

$$M_B = -R_2 \ (6) + 120 + 50 \ (2)$$
$$= - 20 \ \text{kN m}$$

The same result is obtained if we sum the moments to the left of B.

1.6 SHEAR-FORCE AND BENDING-MOMENT DIAGRAMS

Instead of determining shear forces and bending moments at discrete points along a beam it is more convenient for the purposes of beam design to draw diagrams of shear force and bending moment whose ordinates display the variation of these parameters along the beam.

Shear-force and bending-moment diagrams are shown in figure 1.9. They are drawn by considering the shear force and bending moment at a point which starts at the right- (or left-) hand end of the beam.

Suppose the point moves from right to left, then the shear force takes the value + 40 kN until the concentrated load is reached. It then changes to - 10 kN and remains at this value.

The bending moment starts at zero. In bay DE the moment equation is given by

$$M_{DE} = - 40x \ \text{kN m} \qquad\qquad\qquad (a)$$

where x is the distance of the moving point from the reaction R_2.

When D is reached there is a sudden change in moment from - 80 kN m to + 40 kN m due to the presence of the concentrated moment of + 120 kN m.

In bay CD the bending moment is given by

$$M_{CD} = - 40x + 120 \ \text{kN m} \qquad\qquad\qquad (b)$$

and in bay AC

$$M_{AC} = - 40x + 120 + 50 \ (x-4)$$
$$= 10x - 80 \ \text{kN m} \qquad\qquad\qquad (c)$$

The above equations a, b and c give the straight-line segments of the bending moment diagram shown in figure 1.9.

Example 1.4

Draw the bending-moment and shear-force diagrams for the cantilever ABC shown in figure 1.10.

14

The cantilever is 5 m long and is built into a wall at A. It carries a concentrated load of 20 kN at B, 2 m from A and a uniformly distributed load of intensity 10 kN m^{-1} from B to the free end.

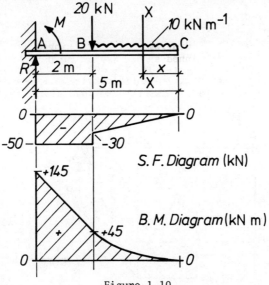

Figure 1.10

The wall provides a reactive force R and a moment M to maintain equilibrium, thus

$$R = 20 + 10(3) = 50 \text{ kN}$$

$$M = 20(2) + 10(3)(3.5) = 145 \text{ kN}$$

The effective line of action of the uniformly distributed load is at the centre of its length which is 3.5 m from A.

To draw the shear-force and bending-moment diagrams consider a section XX distant x from C.

From C to B, $0 < x \leq 3$ m, thus

$$Q_x = -10x \text{ kN}$$

and $M_x = 10x \left(\dfrac{x}{2}\right) = 5x^2 \text{ kN m}$

From B to A, $3 \text{ m} \leq x \leq 5$ m, thus

$$Q_x = -30 - 20 = -50 \text{ kN}$$

and $M_x = 20(x - 3) + 30(x - 1\cdot5)$

$$= 50x - 105 \text{ kN m}$$

The diagrams drawn from these equations are shown in figure 1.10.

15

It should be noted that under the uniformly distributed load the shear-force diagram is described by an inclined straight line and the bending-moment diagram by a parabola.

Example 1.5

Determine the shear-force and bending-moment diagrams for the hinged beam, ABCD, shown in figure 1.11a. A is a pinned support, B and D are roller supports, C is a hinge and the beam is continuous through B. The loading consists of an anticlockwise concentrated moment of 2 kN m at A and a uniformly distributed load of intensity 3 kN m^{-1} on CD.

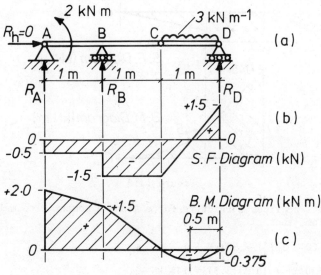

Figure 1.11

At first sight the problem appears to be statically indeterminate since there are four independent reactions. However, this problem has one extra piece of information which will allow us to determine the reactions. Note that in addition to the three equations of statical equilibrium we know that the moment at the hinge must be zero, thus resolving horizontally and vertically

$$R_h = 0$$

$$R_A + R_B + R_C - 3 = 0$$

Taking moments about A and C we have

$$- 2 - 1R_B - 3R_D + 3(2 \cdot 5) = 0$$

and $- 1R_D + 3(0 \cdot 5) = 0$

From these equations the reactions are given by

$$R_A = 0 \cdot 5 \text{ kN}$$

16

$$R_B = 1 \cdot 0 \text{ kN}$$
$$R_D = 1 \cdot 5 \text{ kN}$$

and the shear-force and bending-moment diagrams are as shown in figures 1.11b and c.

1.7 RELATIONS BETWEEN LOAD, SHEAR AND BENDING MOMENT

Consider the equilibrium of the beam element of length δx shown in figure 1.12. The element is at a distance x measured from the left-hand end of the beam. The intensity of loading, w, on the beam is continuously varying. To preserve equilibrium the bending moment and shear force on each face of the element will differ by the small quantities δM and δQ.

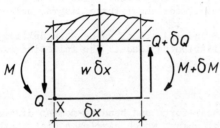

Figure 1.12

The load intensity on the small element is assumed to be constant and the applied load therefore has the value $w \delta x$.

Resolving vertically we have

$$Q + w \, \delta x = Q + \delta Q$$

hence

$$w \, \delta x = \delta Q \tag{a}$$

Taking moments about X we have

$$M + (Q + \delta Q) \, \delta x = M + \delta M + w \, \delta x \left(\frac{\delta x}{2}\right)$$

Ignoring products of small quantities this becomes

$$\delta M = Q \, \delta x \tag{b}$$

As δx becomes vanishingly small, equations a and b may be written in differential form as

$$w = \frac{dQ}{dx} \tag{1.3}$$

and $\quad Q = \dfrac{dM}{dx}$ (1.4)

17

On integrating equation 1.3 between two points 1 and 2 in the beam we obtain

$$Q_2 - Q_1 = \int_{x_1}^{x_2} w \, dx$$

Thus the change in shear force between two points in a beam is equal to the area of the load diagram between these points.

Similarly by integrating equation 1.4

$$M_2 - M_1 = \int_{x_1}^{x_2} Q \, dx$$

Thus the change in bending moment between two points in a beam is equal to the area of the shear-force diagram between these points.

Equations 1.3 and 1.4 are of value when dealing with problems of non-uniformly distributed loads. The following example will illustrate their use.

Example 1.6

Determine the shear-force and bending-moment diagrams for a simply supported beam of span L which carries a distributed load varying

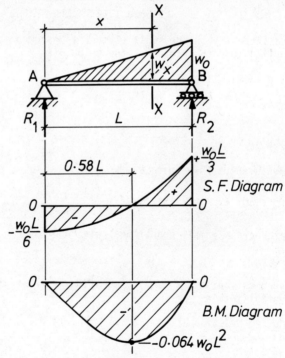

Figure 1.13

18

linearly from an intensity of zero at the left-hand end to w_0 at the right-hand end. Determine also the maximum bending moment and where it occurs.

Simple supports mean that the beam is statically determinate. We may therefore safely assume that one end has a roller support. It is clear in this problem that the horizontal reaction is zero.

Refer to figure 1.13. The load diagram is triangular in shape, so that to determine the reactions R_1 and R_2 the total load may be treated as acting at a point $2L/3$ from A, then resolving vertically

$$R_1 + R_2 = \frac{w_0 L}{2}$$

and taking moments about A

$$R_2 L = \frac{w_0 L}{2} \times \frac{2L}{3}$$

therefore

$$R_2 = \frac{w_0 L}{3} \quad \text{and} \quad R_1 = \frac{w_0 L}{6}$$

At a section XX distant x from A, the ordinate of the load diagram is given by

$$w_x = w_0 \frac{x}{L}$$

The shear force at XX is thus

$$Q_x = \int w_x \, dx = \frac{w_0 x^2}{2L} + A$$

where A is a constant of integration. To determine the constant we note that when $x = L$

$$Q_x = R_2 = \frac{w_0 L}{3}$$

therefore

$$A = \frac{w_0 L}{3} - \frac{w_0 L}{2} = -\frac{w_0 L}{6}$$

and $Q_x = \frac{w_0 L}{2} \left[\left(\frac{x}{L}\right)^2 - \frac{1}{3} \right]$ (1)

The bending moment at XX, is thus

$$M_x = \int Q_x \, dx = \frac{w_0 L}{2} \left(\frac{x^3}{3L^2} - \frac{x}{3} + B \right)$$

where B is another constant of integration.

Since the beam is simply supported, the bending moments at the ends are zero, so that when $x = 0$, $M_x = 0$ and $B = 0$ thus

$$M_x = \frac{w_0 L^2}{6} \left(\frac{x}{L}\right) \left[\left(\frac{x}{L}\right)^2 - 1\right] \qquad (2)$$

From equation 1.4 it is clear that the maximum bending moment will occur when the shear force is zero. If the point of zero shear force is at x_m from A, we have from equation 1 that

$$x_m = \frac{L}{\sqrt{3}} = 0.58L$$

Substituting for x_m in equation 2, the maximum bending moment is given by

$$(M_x)_{max} = -\frac{w_0 L^2}{9\sqrt{3}} = -0 \cdot 064 w_0 L^2$$

The shear-force and bending-moment diagrams are shown in figure 1.13.

1.8   INFLUENCE LINES FOR SHEAR FORCE AND BENDING MOMENT

The shear-force and bending-moment diagrams we have been discussing show the values of these quantities at all points in the beam when the load is fixed in position. The influence lines on the other hand show the variation in shear force or bending moment *at a particular point* as the load moves across the beam.

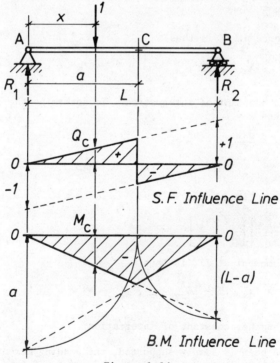

Figure 1.14

20

Consider the simply supported beam AB in figure 1.14 which is subjected to a single concentrated moving load of unit magnitude.

It is required to find the influence lines for bending moment and shear force at the point C, distant a from A.

The reactions R_1 and R_2 are obtained by resolving vertically and taking moments about A, thus

$$R_1 = 1 \times \left(1 - \frac{x}{L}\right)$$

$$R_2 = 1 \times \frac{x}{L}$$

When the unit load is to the left of C, the shear force at C is R_2. When the load is to the right, the shear force is $- R_1$.

Similarly the bending moment at C is $- R_2 (L - a)$ when the load is to the left and $- R_1 a$ when the load is to the right.

Figure 1.14 shows the influence lines for bending moment and shear force at the point C. The ordinates under the load position give the values of these parameters at C.

Note that for a unit dimensionless load the ordinate of the shear-force influence line is dimensionless and the ordinate of the bending-moment influence line has the units of length.

Example 1.7

A lorry weighing 250 kN crosses a simply supported bridge spanning 45 m. The lorry has four wheels and a wheelbase of 5 m. The load on

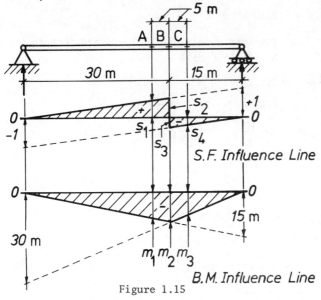

Figure 1.15

21

the rear wheels is 150 kN. If the lorry travels from left to right, determine the bending moment and shear force at a point 15 m from the right-hand end of the bridge when

(a) the front wheels are just to the left of this point
(b) the rear wheels are just to the right of this point.

The shear force at B is obtained from the influence line for shear force by multiplying the appropriate ordinates by the applied loads. Referring to figure 1.15, when the front wheels of the lorry are just to the left of B we have for case (a)

$$Q_B = 150s_1 + 100s_2 \text{ kN}$$

By similar triangles

$$s_1 = \frac{5}{9}, \quad s_2 = \frac{2}{3}$$

therefore

$$Q_B = 150 \left(\frac{5}{9} \right) + 100 \left(\frac{2}{3} \right)$$

$$= + 150 \text{ kN}$$

Similarly for the bending moment at B we have

$$M_B = 150m_1 + 100m_2 \text{ kN m}$$

where $m_1 = -25/3$ m, $m_2 = -10$ m, therefore

$$M_B = -150 \left(\frac{25}{3} \right) - 100(10)$$

$$= -2250 \text{ kN m}$$

(b) When the rear wheels of the lorry are just to the right of B we have by similar reasoning

$$Q_B = 150s_3 + 100s_4 \text{ kN}$$

where $s_3 = -1/3$, $s_4 = -2/9$, therefore

$$Q_B = -150 \left(\frac{1}{3} \right) - 100 \left(\frac{2}{9} \right)$$

$$= -72.2 \text{ kN}$$

also

$$M_B = 150m_2 + 100m_3$$

where $m_3 = -20/3$, therefore

$$M_B = -150(10) - 100 \left(\frac{20}{3} \right)$$

$$= -2167 \text{ kN m}$$

Example 1.8

Two loads W and $2W$ a distance $L/3$ apart cross a simply supported beam of span L. Find the position of the loads that will produce the greatest bending moment in the beam and calculate the magnitude of this moment.

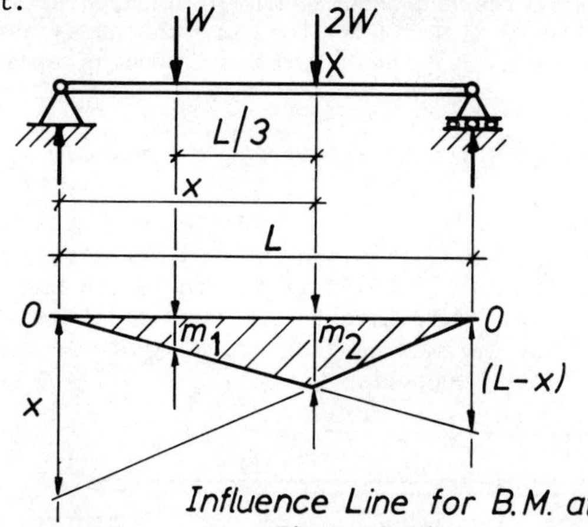

Influence Line for B.M. at X
Figure 1.16

Refer to figure 1.16. Let X be any arbitrary point on the beam. The maximum moment at X will occur when the $2W$ load is at X. Therefore

$$M_x = Wm_1 + 2Wm_2$$

$$\frac{m_1}{x-L/3} = \frac{L-x}{L} \text{ therefore } m_1 = \frac{(3x-L)}{3L}(L-x)$$

$$\frac{m_2}{x} = \frac{L-x}{L} \text{ therefore } m_2 = \frac{x}{L}(L-x)$$

hence

$$M_x = \frac{W(L-x)}{3L}(9x-L)$$

Suppose the maximum value of M_x occurs when $x = x_m$, then

$$\frac{dM_x}{dx} = \frac{W}{3L}\left[(L-x_m)9 - (9x_m-L)\right] = 0$$

hence

$$x_m = \frac{5L}{9}$$

The greatest value of the bending moment in the beam is therefore given by

$$(M_x) = \frac{W}{3L}(L-x_m)(9x_m-L)$$

$$= \frac{16WL}{27} = 0.59\ WL$$

23

The two examples above involved concentrated loads. When the load is distributed there is no difference in the general approach except that we treat the distributed load as an infinite number of concentrated loads and integrate to obtain the final result.

Thus the shear force and bending moment at a point in a beam subjected to a uniformly distributed load are the areas under the load of the influence lines for shear force and bending moment at the point.

Two examples will serve to illustrate the method.

Example 1.9

A simply supported beam of 15 m span is subjected to a uniformly distributed load of 200 kN m^{-1} which is 5 m in length and can occupy any position on the span. Calculate the maximum positive and negative shear-forces and the maximum bending moment that can occur at a point 5 m from the left-hand support.

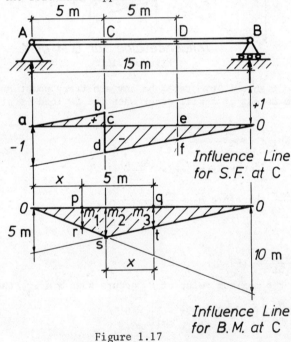

Figure 1.17

There is no difficulty in seeing from figure 1.17 that the maximum positive shear-force at C will occur when the load covers the beam from A to C. The shear-force is given by the area abc (which will be in m) multiplied by the intensity of the load (in kN m^{-1}). Thus the maximum positive shear-force at C

$$= 200 \left(\frac{5}{2} \times \frac{1}{3} \right) \text{ kN}$$

$$= 166 \cdot 7 \text{ kN}$$

24

Similarly the maximum negative shear-force at C which occurs when the load extends from C to D

$$= 200 \times \text{(area cdef)} \text{ kN}$$

$$= 500 \text{ kN}$$

To determine the maximum bending moment at C, let the left-hand end of the load be at x m from A. Then

$$M_C = -200 \times \text{(area pqrst)} \text{ kN m}$$

where

$$\text{area pqrst} = (m_1 + m_2) \frac{(5-x)}{2} + (m_2 + m_3) \frac{x}{2} \text{ m}^2$$

but $\dfrac{m_1}{x} = \dfrac{10}{15}$ or $m_1 = \dfrac{2x}{3}$ m

$$\frac{m_2}{5} = \frac{10}{15} \text{ or } m_2 = \frac{10}{3} \text{ m}$$

and $\dfrac{m_3}{10-x} = \dfrac{5}{15}$ or $m_3 = \dfrac{10-x}{3}$ m

therefore

$$M_C = -\frac{200}{6} (50 + 20x - 3x^2) \text{ kN m}$$

Suppose that the maximum value of the bending moment at C occurs when $x = x_m$ then

$$\frac{d(M_C)}{dx} = \frac{200}{6} (20 - 6x_m) = 0$$

or $\quad x_m = \dfrac{10}{3}$ m

and $(M_C)_{\text{max}} = -\dfrac{200}{6} \left(50 + \dfrac{200}{3} - 3 \times \dfrac{100}{9} \right) = -2778$ kN m

It is interesting to note that if we substitute for x_m in the expressions for m_1 and m_3 we find that

$$m_1 = m_3 = \frac{20}{9} \text{ m}$$

The equality of m_1 and m_3 provides a simple rule for determining the maximum bending moment at a point for distributed loads wholly on the span.

Example 1.10

Sketch the influence lines for vertical and moment reaction at the built-in end A of the beam system ABCDEF shown in figure 1.18a. B and D are hinges and C and E are roller supports. Determine the

25

maximum moment at A when the beam is crossed by a uniformly distributed load of length 5 m and intensity 10 kN m^{-1}.

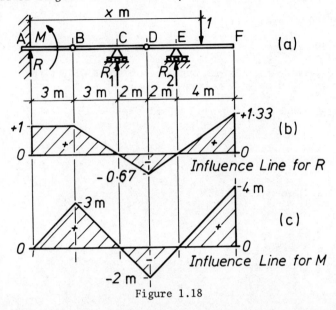

Figure 1.18

Let the positive directions for M and R be as shown in figure
1.18a. Suppose now that a unit load crosses the beam from right to
left and let the distance of this load from A be x m.

Suppose $x > 8$ m, then resolving vertically we have

$$R + R_1 + R_2 = 1 \qquad (1)$$

and summing moments about A and the hinges D and B we have

$$M + 6R_1 + 10R_2 = x \times 1 \qquad (2)$$

$$2R_2 = 1(x-8) \qquad (3)$$

$$3R_1 + 7R_2 = 1(x-3) \qquad (4)$$

From equations 1, 2, 3 and 4 we have for $x > 8$ m

$$R = \frac{(x-10)}{3}$$

$$M = (x - 10) \text{ m} \qquad (5)$$

For $3 < x < 8$ m equations 1, 2 and 4 apply, but summing moments about
D yields $R_2 = 0$, thus

$$R = \frac{(6-x)}{3}$$

$$M = (6-x) \text{ m} \qquad (6)$$

26

For $x < 3$ m equations 1 and 2 apply, but summing moments about B and D yields $R_1 = R_2 = 0$, thus

$$R = 1$$

$$M = x \text{ m} \qquad\qquad (7)$$

The influence lines for R and M shown in figure 1.18 are drawn from equations 5, 6 and 7.

When the distributed load crosses the span, the greatest value of M will occur when the area under a 5 m length of the influence line for M is a maximum. By inspection, this will occur when the load is symmetrically placed about the hinge B (see figure 1.19).

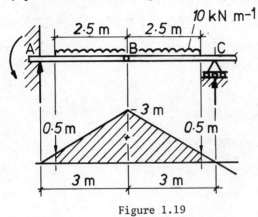

Figure 1.19

The maximum area under the influence line for M is $(3 \cdot 5)$ $(2 \cdot 5) = 8 \cdot 75$ m^2. Thus for a distributed load of intensity 10 kN m^{-1}

$$M_{max} = (10) \ (8.75)$$

$$= 87 \cdot 5 \text{ kN m}$$

1.9 THE THREE-PINNED ARCH

Because of its curved shape, an arch under load develops axial compressive forces as well as shear forces and bending moments.

The points where the arch is supported are called abutments. If the arch is pinned at the abutments, four independent reactions will be generated (one horizontal and one vertical at each abutment). The structure is therefore statically indeterminate since the three equations of statical equilibrium are insufficient to complete the force analysis. To make the structure statically determinate we require a third pin which is usually inserted at the crown (the highest point in the arch).

Arches may be of any shape but the most common are parabolic or circular.

Example 1.11

Determine the reactions for the unsymmetrical three-pinned parabolic arch shown in figure 1.20. Find also the bending moment at a point D, 6·5 m to the left of the crown. The load may be assumed to be uniformly distributed horizontally.

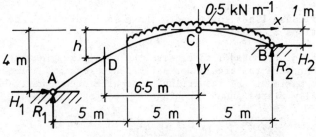

Figure 1.20

Resolving horizontally

$$H_1 - H_2 = 0 \tag{1}$$

Resolving vertically

$$R_1 + R_2 = 10(0·5) = 5 \text{ kN} \tag{2}$$

Summing moments about the abutment A

$$15R_2 + 3H_2 - 50 = 0 \tag{3}$$

Summing moments to the right of the crown C

$$5R_2 - H_2 - 6.25 = 0 \tag{4}$$

From equations 3 and 4

$$R_2 = 2·29 \text{ kN and } H_2 = 5·2 \text{ kN}$$

From equations 1 and 2

$$R_1 = 2·71 \text{ kN and } H_1 = 5·2 \text{ kN}$$

Since $H_1 = H_2$, the horizontal component of thrust in the arch is constant.

The equation describing the parabolic shape of the arch is $y = ax^2$ if the axes are as shown in figure 1.20 with the origin at the crown C. Since the pin B has coordinates $x = 5$ m and $y = 1$ m the constant a is given by

$$a = \frac{1}{5^2} = \frac{1}{25} \text{ m}^{-1}$$

This result may be used to determine the vertical displacement of

28

point D with respect to the crown, thus

$$h = \frac{1}{25} (6 \cdot 5)^2 = 1 \cdot 69 \text{ m}$$

Summing moments to the left of D we have

$$M_D = - 3 \cdot 5 R_1 + (4-h) H_1$$

$$= - 3 \cdot 5 (2 \cdot 71) + 2 \cdot 31 (5 \cdot 2)$$

$$= + 2 \cdot 52 \text{ kN m}$$

The positive sign indicates that the moment is hogging.

Example 1.12

The three-pinned arch shown in figure 1.21 is in the form of an arc of a circle of radius 15 m. Determine the reactions at the abutments and the bending moment at a point D, 3 m to the left of the crown.

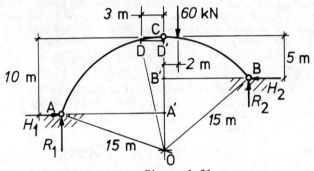

Figure 1.21

Resolving horizontally and vertically we have

$$R_1 + R_2 = 60 \text{ kN} \tag{1}$$

and $H_1 - H_2 = 0$ (2)

Summing moments about A and C, we have

$$R_2(10\sqrt{2} + 5\sqrt{5}) + 5H_2 = 60(10\sqrt{2} + 2) \text{ kN m} \tag{3}$$

and $R_2(5\sqrt{5}) - 5H_2 = 60(2) \text{ kN m}$ (4)

From equations 3 and 4

$$R_2 = 29 \cdot 8 \text{ kN and } H_2 = 42 \cdot 7 \text{ kN}$$

Hence from equations 1 and 2

$$R_1 = 30 \cdot 2 \text{ kN and } H_1 = 42 \cdot 7 \text{ kN}$$

29

The bending moment at D is given by

$$M_D = - R_1(10\sqrt{2} - 3) + H_1(A'D')$$

Now

$$A'D' = OD' - OA' = \sqrt{(15^2 - 3^2)} - 5 = 9 \cdot 7 \text{ m}$$

Thus

$$M_D = - 336 \cdot 3 + 413 \cdot 9 = + 77 \cdot 6 \text{ kN m}$$

1.10 SUSPENSION CABLES

A suspension cable looks rather like an inverted arch, but unlike
the arch, which is subject to compression and bending, a flexible
cable carries loads by the development of axial tension alone.

Suspension cables have their most obvious use in suspension
bridges. The cable is hung between two towers and is connected with
the bridge deck by a large number of vertical hangers. If the bridge
deck is also partly supported at the points where it passes through
the towers, the problem is statically indeterminate and the force
analysis becomes quite complex. We shall be concerned here only with
the statically determinate type of problem in which the whole of the
loading is carried by the cable.

The main assumption of the simple cable theory is that the sag is
small compared with the span. We are then justified in taking the
self-weight to be uniformly distributed horizontally. The cable is
also assumed to be inextensible.

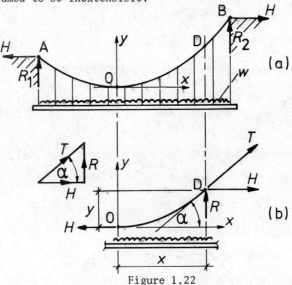

Figure 1.22

The suspension cable AOB shown in figure 1.22a carries a uniformly

30

distributed load (on the horizontal) of intensity w per unit length which includes the cable self-weight. Since there are no horizontally applied loads, the horizontal component of the cable tension is constant and equal to H.

We now consider the equilibrium of the length of cable OD shown in figure 1.22b. At D the cable tension is T and the vertical and horizontal components of T are R and H respectively. At O, the lowest point of the cable, the cable tension is horizontal and equal to H. There is no vertical component at O. The horizontal equilibrium of OD is assured since the horizontal component of the cable tension is constant. For vertical equilibrium we require that

$$R = wx \tag{a}$$

From the triangle of forces at D we have

$$\tan \alpha = \frac{R}{H} = \frac{wx}{H} \tag{b}$$

But the slope of the cable curve at A is given by

$$\frac{dy}{dx} = \tan \alpha \tag{c}$$

From equations b and c therefore ,

$$\frac{dy}{dx} = \frac{wx}{H} \tag{d}$$

Integrating equation d and noting that $y = 0$ where $x = 0$ we have

$$y = \frac{wx^2}{2H} \tag{1.5}$$

Equation 1.5 is the equation of a parabola. Had we carried out the exact analysis in which the cable self-weight is distributed uniformly along its *length* the corresponding equation would have been that of a catenary.

1.10.1 Reactions on Suspension-bridge Towers

(a) Probably the most common method of supporting a suspension cable at the tower is by means of a saddle over which the cable passes. The saddle in turn rests on rollers. Figure 1.23 shows the arrangement.

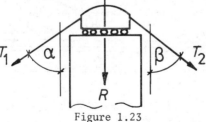

Figure 1.23

The main load-carrying length of cable is on one side of the tower.

On the other side the cable is led down to anchorage points. Since the saddle is mounted on rollers, the horizontal components of the cable tensions on each side of the tower must be equal thus

$$T_1 \sin \alpha = T_2 \sin \beta \tag{a}$$

The reaction on the tower is therefore vertical and is given by

$$R = T_1 \cos \alpha + T_2 \cos \beta \tag{b}$$

(b) An alternative way of supporting the cable is to pass it over a pulley fixed to the top of the tower as shown in figure 1.24.

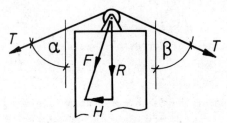

Figure 1.24

If the pulley is assumed to be frictionless, the tension in the cable will remain unaltered as it passes over. If α is not equal to β the reaction F on the tower will have vertical and horizontal components R and H respectively. Referring to figure 1.24 we see that

$$R = T (\cos \alpha + \cos \beta) \tag{c}$$

and $H = T (\sin \alpha - \sin \beta)$ $\tag{d}$

Example 1.13

A suspension cable AB has a span of 200 m, end B being 3 m above A and the lowest point being 16 m below A. If the loading on the cable

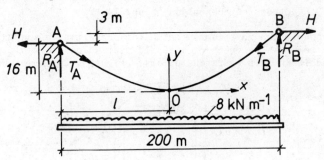

Figure 1.25

has an intensity of 8 kN m^{-1}, calculate the vertical and horizontal loads on the towers. The anchor cables at each end make an angle of

32

22° with the horizontal. The cable passes over a pulley at A and a saddle on rollers at B.

Refer to figure 1.25. Let O, the lowest point in the cable, be ℓ m from A. Applying equation 1.5 to each side of O we have

$$16 = \frac{4\ell^2}{H} \tag{1}$$

and $$19 = \frac{4}{H}(200-\ell)^2 \tag{2}$$

From equations 1 and 2 we obtain

$$\ell = 95 \cdot 7 \text{ m}$$

and $$H = 2290 \text{ kN}$$

The load on the cable between A and O must be supported by the reaction at A, thus

$$R_A = (8)(95 \cdot 7) = 765 \cdot 6 \text{ kN}$$

hence

$$R_B = (8)(200) - 765 \cdot 6 = 834 \cdot 4 \text{ kN}$$

From the triangle of forces at the supports we have

$$T_A = \sqrt{(H^2 + R_A^2)} = 2415 \text{ kN}$$

and $$T_B = \sqrt{(H^2 + R_B^2)} = 2437 \text{ kN}$$

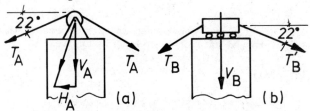

Figure 1.26

Figure 1.26 shows the cable-support arrangements. At end A (figure 1.26a) the cable tension is constant over the pulley thus the tower reactions are

$$V_A = R_A + T_A \sin 22° = 1670 \text{ kN}$$

and $$H_A = H - T_A \cos 22° = 51 \cdot 2 \text{ kN}$$

At end B, (figure 1.26b) the horizontal components of the cable tensions are equal, thus

$$H = T_B' \cos 22°$$

33

Hence

$$T_B' = 2470 \text{ kN}$$

The reaction on the tower is vertical and is given by

$$V_B = R_B + T_B' \sin 22° = 1760 \text{ kN}$$

Example 1.14

A uniform inextensible cable of weight W is suspended with a small sag d_0 between two fixed points at the same level. A point load of $3W$ is then attached to the centre of the cable. Estimate the new value of the sag at the centre.

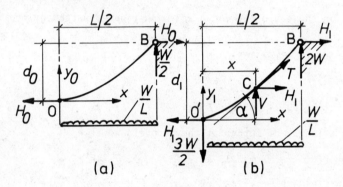

Figure 1.27

Figures 1.27a and b show the forces and displacements in one half of the cable before and after the $3W$ load is applied.

Before loading, the cable equation is given by

$$y_0 = \frac{W}{L} \frac{x^2}{2H_0} \qquad (1)$$

But $y_0 = d_0$ when $x = L/2$, thus

$$d_0 = \frac{WL}{8H_0}$$

hence

$$y_0 = 4d_0 \left(\frac{x}{L}\right)^2 \qquad (2)$$

At a point C in the cable after loading, the vertical and horizontal components of the cable tension are V and H_1, where

$$V = \frac{3W}{2} + \left(\frac{W}{L}\right) x$$

34

If the tangent at C is inclined at an angle α to the horizontal, we have

$$\tan \alpha = \frac{V}{H_1} = \frac{dy_1}{dx}$$

or $\quad \dfrac{dy_1}{dx} = \dfrac{3W}{2H_1} + \dfrac{Wx}{LH_1}$

Integrating, and noting that $y_1 = 0$ when $x = 0$ we obtain

$$y_1 = \frac{3Wx}{2H_1} + \frac{Wx^2}{2LH_1} \tag{3}$$

Now $y_1 = d_1$ when $x = L/2$, thus

$$d_1 = \frac{7WL}{8H_1}$$

or $\quad H_1 = \dfrac{7WL}{8d_1}$

The equation of the loaded cable curve is therefore obtained by substituting for H_1 in equation 3 thus

$$y_1 = \frac{4d_1}{7} \left[3\left(\frac{x}{L}\right) + \left(\frac{x}{L}\right)^2 \right] \tag{4}$$

Since the cable is inextensible, the cable lengths before and after loading are the same.

Now the length, ds, of a small element of the cable is given by

$$ds = \surd(dx^2 + dy^2) = dx \left[1 + \left(\frac{dy}{dx}\right)^2 \right]^{1/2}$$

Expanding the root and ignoring powers of (dy/dx) greater than the second we have

$$ds \simeq \left[1 + \frac{1}{2}\left(\frac{dy}{dx}\right)^2 \right] dx$$

The total length of cable, S, is then given by

$$S = 2\int_0^{L/2} ds = 2\int_0^{L/2} \left(1 + \frac{1}{2}\left(\frac{dy}{dx}\right)^2 \right) dx$$

If the cable is to be inextensible, we require that

$$\int_0^{L/2} \left(\frac{dy_0}{dx}\right)^2 dx = \int_0^{L/2} \left(\frac{dy_1}{dx}\right)^2 dx \tag{5}$$

Now $\quad \dfrac{dy_0}{dx} = \dfrac{8d_0}{L^2} x$

and $\quad \dfrac{dy_1}{dx} = \dfrac{4d_1}{7L^2}(3L + 2x)$

35

thus substituting in equation 5 from equations 2 and 4 we have

$$\frac{64d_0{}^2}{L^4} \int_0^{L/2} x^2 dx = \frac{16d_1{}^2}{49L^4} \int_0^{L/2} (9L^2+12xL+4x^2) \ dx$$

hence

$$d_1 = \frac{7d_0}{\sqrt{37}} = 1 \cdot 15 \ d_0$$

The next example deals with a similar configuration but this time only forces are required and the solution is much simpler.

Example 1.15

A concentrated load of 350 N is supported by a flexible suspension cable of weight 7 N m^{-1} in such a manner that the load point is 3 m from the left-hand abutment and 1·5 m below it. The cable supports are 12 m apart and the right-hand end of the cable is 4·5 m above the left-hand end. Determine the maximum cable-tension and where it occurs. The cable weight may be taken as uniformly distributed horizontally.

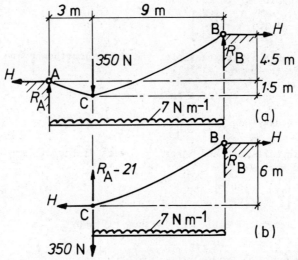

Figure 1.28

Figure 1.28a shows the forces acting on the whole cable, while figure 1.28b shows the forces on the length CB.

Resolving forces vertically for the whole cable we have

$$R_A + R_B = 350 + 84 = 434 \text{ N} \tag{1}$$

Taking moments about A

$$3(350) + 4 \cdot 5(H) + \frac{12^2}{2} (7) = R_B(12) \text{ N m}$$

hence

$$8R_B - 3H = 1036 \text{ N} \tag{2}$$

Taking moments about C for the length of cable CB (figure 1.28b) we have

$$6(H) + \frac{9^2}{2}(7) = 9R_B \text{ N m}$$

hence

$$3R_B - 2H = 94 \cdot 5 \text{ N} \tag{3}$$

From equations 2 and 3

$$H = 336 \text{ N and } R_B = 255 \cdot 5 \text{ N}$$

From equation 1

$$R_A = 178 \cdot 5 \text{ N}$$

By inspection, the maximum cable tension, T_m, is at B and is given by

$$T_m = \sqrt{[(336)^2 + (255 \cdot 5)^2]} = 422 \text{ N}$$

1.11 PROBLEMS FOR SOLUTION

1. A heavy uniform rod of weight W is hung from a point by two equal strings, one attached to each end of the rod. A body of weight w is hung halfway between the centre and one end of the rod. Prove that the ratio of the string tensions is $(2W+3w)/(2W+w)$.

2. Determine the single resultant force that can replace the system in figure 1.29 and find also the intersections of its line of action with AB and CB.
($9 \cdot 16$ kN; $0 \cdot 47$ m from B on AB, $1 \cdot 35$ m from B on CB)

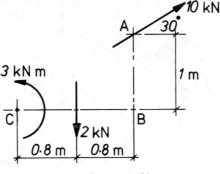

Figure 1.29

3. Determine the forces in the members of the tripod wall bracket in figure 1.30.
($71 \cdot 9$, $-25 \cdot 6$, $+129 \cdot 0$ kN)

37

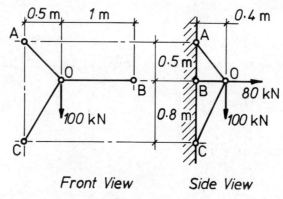

0.5 m 1 m 0.4 m

A

0.5 m

O

B 80 kN

100 kN 0.8 m 100 kN

C C

Front View   Side View

Figure 1.30

4. For the simply supported Warren girder shown in figure 1.31 determine the value of *W* if the maximum bar tension is 60 kN. All bars
are the same length. What is the maximum compression and in which
bar does it occur?
(17·32 kN, 60 kN)

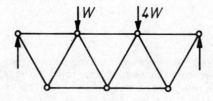

↓W   ↓4W

Figure 1.31

5. Determine the largest compressive and tensile forces in the frame
shown in figure 1.32.
(+17·3, -28·4 kN)

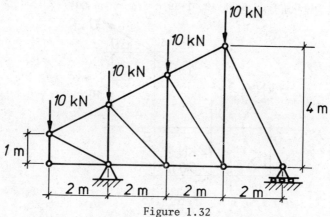

↓10 kN

↓10 kN

↓10 kN

↓10 kN

4 m

1 m

2 m   2 m   2 m   2 m

Figure 1.32

6. Determine the forces in the tripod shown in figure 1.33.
(+8·73, -34·40, +5·64 kN)

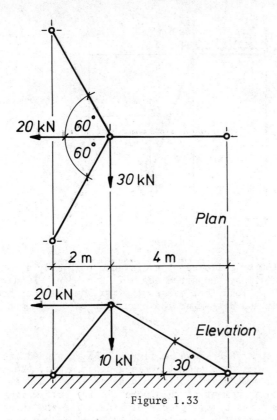

20 kN   60°

60°

30 kN

Plan

2 m   4 m

20 kN

Elevation

10 kN   30°

Figure 1.33

7. Determine the shear-force and bending-moment diagrams for a beam simply supported at its ends for which the loading is given by

$$w = w_m \sin \frac{\pi x}{L}$$

The origin of the x-axis is at the left-hand end of the beam and L is the beam length.

$$\left(\text{S.F.} = -\frac{w_m L}{\pi} \cos \frac{\pi x}{L}, \ \text{B.M.} = -\frac{w_m L^2}{\pi^2} \sin \frac{\pi x}{L} \right)$$

8. A beam supporting a uniformly distributed load over its entire length L is pinned at one end and passes over a roller support at a distance a from the other end. Determine the value of a such that the greatest bending moment in the beam is as small as possible.
$(0 \cdot 293L)$

9. The frame ABCD shown in figure 1.34 is pinned at A and is attached to a roller support at D. The loading consists of a uniformly distributed load of 5 kN m^{-1} from A to B, a concentrated vertical load of 10 kN at B and a concentrated horizontal load of 12 kN at C, the mid-point of BD. Draw the bending-moment and shear-force diagrams for the frame and state maximum values.
$(-26 \cdot 9 \text{ kN}, \ -72 \cdot 2 \text{ kN m})$

39

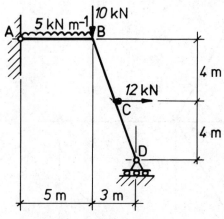

Figure 1.34

10. A load of 100 kN followed at a distance of 8 m by a load of 50 kN passes across a simply supported beam of 20 m span. Find the position of the 100 kN load that causes the maximum value of bending moment in the beam and determine the magnitude of this moment. (-563·3 kN m)

11. The beam system ABCD shown in figure 1.35 is pinned at A and supported on rollers at B and D. C is a hinge. Draw the influence line for bending moment at a point midway between A and B. Determine the maximum bending moment at this point if the self-weight of the beam is 10 kN m^{-1} and a moving load of length 2 m and intensity 20 kN m^{-1} passes over the beam. (-65 kN m)

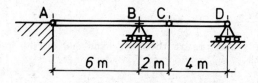

Figure 1.35

12. A circular arch of span 50 m is pinned at the abutments which are at the same level. A third pin is at the crown which is 5 m above the abutments. The arch carries a uniformly distributed load of 30 kN m^{-1} over the whole span and an additional uniformly distributed load of 15 kN m^{-1} over the left-hand half of the span. Determine the horizontal and vertical reactions at the abutments and the bending moment at the left-hand quarter-span point. Assume that the loads are uniformly distributed horizontally. (2344, 1031 and 844 kN; -500 kN m)

13. Draw the influence line for horizontal thrust for the three-pinned parabolic arch ACB shown in figure 1.36. Hence determine the

40

maximum thrust when a uniformly distributed load of intensity 50 kN m^{-1} and length 10 m passes over the span.
(277·8 kN)

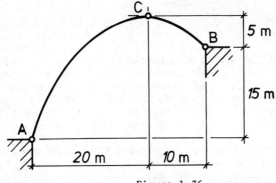

Figure 1.36

14. A suspension cable is hung between two points distance L apart and at the same height. Under load the centre of the cable sags a distance d below the supports. If the vertical loading on the cable is given by

$$w = w_m \sin \frac{\pi x}{L}$$

where x is measured from one end, determine the maximum value of the cable tension.
$$\frac{w_m L}{\pi} \left[1 + \left(\frac{L}{\pi d} \right)^2 \right]^{\frac{1}{2}}$$

15. A twin-cable suspension-bridge with a flexible deck spans 50 m. At one end the cables are secured to an anchorage at a point 11 m above the deck, which is 2 m below the cables' lowest point. At the other end, the cables pass over a tower whose top is 6 m above the deck. The cables pass over the tower on a saddle which rests on rollers and down to anchorage points at an angle of 30° to the horizontal. If the deck loading and cable self-weight is uniformly distributed horizontally over the whole span with an intensity of 12 kN m^{-1} determine the maximum tension in each cable, and the reaction on the tower.
(350 kN, 586 kN)

2 THE STRESS-STRAIN RELATIONSHIP

The intensity at which a force is distributed over the cross-section
of a structural member is referred to as stress. The magnitude and
distribution of the stress will depend upon the way in which the
loads are applied and on the geometric properties of the cross-
section. In general, a member may be required to sustain axial
forces, bending moments, shear forces and torques taken individually
or in combination with each other. In this chapter however we shall
be concerned with stresses due to axial forces alone.

The action of a force on a deformable body is to cause it to
change its shape. In general terms this shape change is expressed by
reference to the strain in the body.

2.1 NORMAL STRESS AND STRAIN

Consider a straight, uniform bar of cross-sectional area A and length
L (figure 2.1a) which carries an axial tensile force P. If it were
possible to look at the surface of the bar exposed by an imaginary
cut XX (figure 2.1b) normal to the bar axis, we should find that the
force P was distributed over the cut surface at a mean intensity, or
stress, of σ (sigma). To preserve longitudinal equilibrium for the
cut bar we find that

$$P = \sigma A$$

or $\quad \sigma = \dfrac{P}{A}$

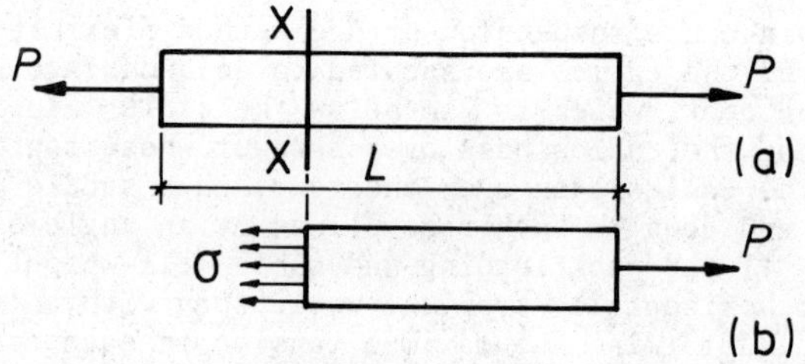

Figure 2.1

A more rigorous definition of stress is obtained if we consider a
small cross-sectional element of area δa, subject to a normal force
δp. Then the stress, σ at the element is defined as

$$\sigma = \lim_{\delta a \to 0} \frac{\delta p}{\delta a} \tag{2.1}$$

This argument also applies to forces tending to compress the bar.
We speak of tensile forces giving rise to tensile stresses and compressive forces giving rise to compressive stresses.

The units of stress depend on the units chosen for the force P and the cross-sectional area A. If, for example, P is in newtons (N) and A in square millimetres (mm^2), the stress will be in N mm^{-2}. Note that the same numerical value will be obtained for the stress if P is measured in meganewtons ($MN = N \times 10^6$) and A in square metres ($m^2 = mm^2 \times 10^6$). Thus the units of stress could equally well be written as $MN \, m^{-2}$ or $N \, mm^{-2}$.

It will readily be appreciated that the tensile forces acting on the bar in figure 2.1 will cause it to stretch or increase its length.

If compressive forces had been applied, the bar length would have shortened. Suppose the extension (or contraction) of the bar is x.

The deformation may be made non-dimensional by dividing by the unstressed length of the bar, L. This non-dimensional deformation is referred to as the strain and is given the symbol ε (epsilon), thus

$$\varepsilon = \frac{x}{L} \qquad\qquad (2.2)$$

2.2   THE STRESS-STRAIN RELATIONSHIP

A very important performance characteristic for a material is its tensile stress-strain curve. This curve may be plotted from the results of a tensile test in which the extension of a sample of the material is measured under gradually increasing tensile force. Stress is calculated by dividing the force by the original cross-sectional area of the sample. Strain is calculated by dividing the extension of a given length (gauge length) by the original unstressed length. Figure 2.2 illustrates the tensile stress-strain curves for some common constructional materials.

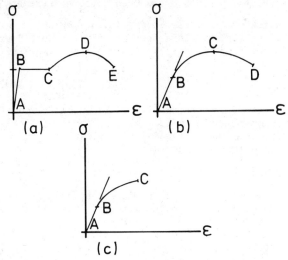

Figure 2.2

The curves in figure 2.2 are not to scale, they are intended merely

43

for comparison purposes. Figure 2.2a shows the typical behaviour of annealed mild-steel. Stress is proportional to strain from A to B. From B to C a large amount of strain occurs at constant stress. This phenomenon, which is peculiar to annealed mild-steel, is known as yielding and the stress at B where yielding starts is called the yield stress (σ_y). In fact the yield stress at B has upper and lower values. The difference between these two is small and the quoted yield stress is usually the lower value. At C the material becomes strain hardened and an increase in stress is required to take the sample to its maximum stress at D. Fracture of the sample occurs at E. The apparent fall in stress from D to E is because the stress is calculated by dividing the axial force by the original cross-sectional area. In fact the longitudinal extension of the sample is accompanied by a reduction in cross-sectional area. This reduction becomes significant near the ultimate stress and if the true stress were plotted it would be seen to continue to rise until fracture occurs.

The shape of the curve in figure 2.2b is characteristic of structural aluminium-alloys and certain high-strength steels. Stress is proportional to strain from A to B, the ultimate stress is at C and fracture occurs at D. The region of large strain at constant stress typical of annealed mild-steel is seen to be absent.

Figure 2.2c shows the type of tensile stress-strain curve obtained for brittle materials such as cast iron or concrete. Stress is approximately proportional to strain for the initial part of the curve from A to B. Fracture occurs at C. The curve at this point is still rising since brittle materials are not capable of sustaining a significant reduction in cross-sectional area.

Stress-strain curves may also be obtained for compressive loading. For most metals the elastic behaviour is similar to that in tension. Brittle materials, however, show much improved strength in compression.

A common feature of these curves is the region in which stress is proportional to strain. This fact was first observed experimentally by Robert Hooke (1635-1703). We express this simple relationship in the form

$$\sigma = E\varepsilon \tag{2.3}$$

The constant of proportionality is known as the modulus of elasticity or Young's modulus after the English scientist Thomas Young (1773-1829). Note that the modulus of elasticity has the same units as stress since strain is non-dimensional.

Equation 2.3 is valid provided the material remains linearly elastic. A material is said to be elastic if, on unloading from a certain stress, it regains the shape it had before the load was applied. Internal breakdown occurs in the material if it is stressed beyond the limit of elasticity. Subsequent unloading then reveals a permanent deformation or set. For mild steel, the limit of elasticity may be assumed to coincide with the yield stress.

Some materials such as rubbers can be elastic without obeying a linear relationship between stress and strain. These are said to be non-linear elastic materials. In this chapter we shall be concerned only with elastic behaviour of materials following a linear relationship between stress and strain. Fortunately this restriction is not serious since all the common constructional materials behave in this way under working loads.

Example 2.1

Determine the greatest length of mild-steel wire of uniform cross-section that may be suspended vertically if the maximum stress is not to exceed the yield stress of 250 MN m^{-2}. The density, ρ, of steel is 7·84 Mg m^{-3} and g, the acceleration due to gravity, is 9·81 m s^{-2}.

The maximum tensile-force, F_m, in the wire occurs at the suspension point and it is equal to the total weight of the suspended wire. Let the length of wire be L m and its cross-sectional area be A m^2. The total volume of wire is $A \times L$ m^3 and its mass is $\rho \times A \times L$ Mg, thus

$$F_m = \rho g A L \text{ Mg m s}^{-2}$$

but 1 Mg m s^{-2} = 1 kN

thus F_m = (7·84)(9·81)AL = 76·91 AL kN

If σ_m is the maximum stress, we have

$$\sigma_m = \frac{F_m}{A} = 76\text{·}91L \text{ kN m}^{-2}$$

but σ_m is not to exceed 250 MN m^{-2}, hence the maximum value of L is given by

$$76\text{·}91L \times 10^{-3} = 250 \text{ MN m}^{-2}$$

or $L = \dfrac{250 \times 10^3}{76\text{·}91} = 3250$ m = 3·25 km

Note that this length is independent of the cross-sectional area.

Example 2.2

A circular-section bar of length L and diameter d has the two end-thirds of its length turned down to a diameter $d/2$. Determine an

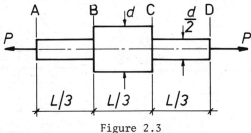

Figure 2.3

45

expression for the elongation of this stepped bar under an axial tensile force P. What is the elongation for a uniform bar of the same material having the same length and volume? Assume that stresses are uniformly distributed across the bar.

Referring to figure 2.3 we have

$$\sigma_{AB} = \sigma_{CD} = \frac{16P}{\pi d^2}$$

and $\sigma_{BC} = \frac{4P}{\pi d^2}$

Dividing by Young's modulus to determine the strains we have

$$\varepsilon_{AB} = \varepsilon_{CD} = \frac{16P}{\pi d^2 E}$$

and $\varepsilon_{BC} = \frac{4P}{\pi d^2 E}$

Thus the total elongation, Δ, is given by

$$\Delta = \frac{2L}{3} \frac{16P}{\pi d^2 E} + \frac{L}{3} \frac{4P}{\pi d^2 E} = \frac{12PL}{\pi d^2 E}$$

The cross-sectional area, A, of a uniform bar of the same volume and length is given by

$$A = \frac{1}{L} \left[\frac{\pi}{4} \left(\frac{d}{2} \right)^2 \frac{2L}{3} + \frac{\pi d^2}{4} \frac{L}{3} \right] = \frac{\pi d^2}{8}$$

The resulting strain is therefore

$$\varepsilon = \frac{8P}{\pi d^2 E}$$

and the associated elongation

$$\Delta' = \frac{8PL}{\pi d^2 E} = \frac{2\Delta}{3}$$

Example 2.3

A straight uniform steel rod of length 60 cm rotates about an axis through one end perpendicular to its length. Estimate the speed of

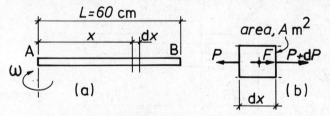

Figure 2.4

rotation that will produce a maximum tensile stress of 80 MN m^{-2} in the rod. What is the total elongation at this speed? The density, ρ, of steel is 7·84 Mg m^{-3} and E is 210 GN m^{-2}.

Refer to figure 2.4. The tensile force in the rod opposes the centrifugal force induced by the rotation. This force will be a maximum at A (the axis of rotation) and zero at B. Consider the element of bar shown in figure 2.4b and let the cross-sectional area of the bar be A m^2. The force equilibrium of the element requires that

$$P + dP + F = P$$

or $F = -dP$ (1)

where F is the centrifugal force on the element. If the mass of the element is m and the angular velocity of the rod is ω rad s^{-1}, the centrifugal force F at a point x from the axis of rotation is given by

$$F = m\omega^2 x$$ (2)

but $m = A\rho\ dx$ (3)

thus from equations 1, 2 and 3 we have

$$dP = -A\rho\omega^2 x\ dx$$

On integrating

$$P = -A\rho\omega^2 \left(\frac{x^2}{2} + K\right)$$

The constant of integration, K, is obtained by noting that $P = 0$ when $x = L$, then

$$K = -\frac{L^2}{2}$$

thus $P = \dfrac{A\rho\omega^2}{2} (L^2 - x^2)$

and the stress in the rod is

$$\sigma = \frac{\rho\omega^2}{2} (L^2 - x^2)$$ (4)

The maximum stress, σ_m, is limited to 80 MN m^{-2} and occurs at A where $x = 0$, thus from equation 4

$$\omega = \frac{1}{L} \sqrt{\left(\frac{2\sigma_m}{\rho}\right)} = \frac{1}{0\cdot6} \sqrt{\left(\frac{2 \times 80 \times 10^3}{7\cdot84}\right)} \text{ rad s}^{-1}$$

hence

$$\omega = 238 \cdot 1 \text{ rad s}^{-1} \quad (37 \cdot 9 \text{ rev s}^{-1})$$

From equation 4 the strain ε at the element is given by

$$\varepsilon = \frac{\rho\omega^2}{2E} (L^2 - x^2)$$

thus the elongation, $d\Delta$, of the element is obtained from

$$d\Delta = \varepsilon \, dx = \frac{\rho\omega^2}{2E} (L^2 - x^2) \, dx \qquad (5)$$

and the total elongation is found by integrating equation 5 over the whole length of the rod, thus

$$\Delta = \frac{\rho\omega^2}{2E} \int_0^L (L^2 - x^2) \, dx = \frac{\rho\omega^2 L^3}{3E}$$

$$= \sigma_m \frac{2L}{3E}$$

hence

$$\Delta = \frac{80 \times 2(0 \cdot 6)}{3 \times 210 \times 10^3} \text{ m}$$

$$= 0 \cdot 152 \text{ mm}$$

2.3 POISSON'S RATIO

We have seen that the effect of the axial load P on the bar in figure 2.1 is to produce an elongation in the axial direction. Had we made measurements of the bar cross-section we would also have found a contraction in the direction at right angles to the bar axis. If we denote the longitudinal strain by ε_x, the lateral strain ε_y is given by

$$\varepsilon_y = -\upsilon \, \varepsilon_x \qquad (2.4)$$

where υ, the ratio of the lateral strain to longitudinal strain, is called Poisson's ratio after the French mathematician Simeon Poisson (1781-1840) who postulated this relationship. The negative sign indicates that a longitudinal elongation is accompanied by a lateral contraction and vice versa. For most metals, υ takes a value between $0 \cdot 25$ and $0 \cdot 35$. For rubber, υ is close to the maximum theoretical value of $0 \cdot 5$.

Example 2.4

Determine the change in volume of a square-section bar of side a and length L under an axial tensile load P.

Let the longitudinal and lateral strains in the bar be ε_x and ε_y respectively, then the new volume of the bar is given by

$$a^2L + \Delta V = [a(1 + \varepsilon_y)]^2 \times L(1 + \varepsilon_x) \qquad (1)$$

where ΔV is the change in volume.

Expanding the right-hand side of equation 1 and neglecting products of strains, we have

$$a^2L + \Delta V = a^2L(1 + \varepsilon_x + 2\varepsilon_y)$$

or

$$\Delta V = a^2L(\varepsilon_x + 2\varepsilon_y)$$

Substituting from equation 2.4 for ε_y, we have

$$\Delta V = a^2L\varepsilon_x(1 - 2\nu)$$

but $\varepsilon_x = \dfrac{P}{a^2E}$

thus

$$\Delta V = \frac{PL}{E}(1 - 2\nu)$$

2.4 THIN CYLINDERS

An example of a simple problem that involves stresses in two directions at right angles is the thin-walled hollow cylinder under internal pressure. Suppose that the cylinder has closed ends and that the internal pressure is p. The mean diameter is d and the wall thickness (which is small compared with d) is t.

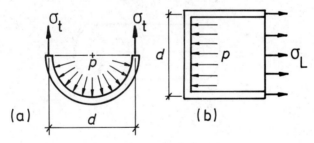

Figure 2.5

Consider the half transverse section of the cylinder away from the ends which is shown in figure 2.5a. This section is formed by cutting at right angles to the cylinder axis and then slicing longitudinally in a plane that includes the cylinder axis. If this section has unit length, the downward force produced by the internal pressure acting on the projected cylinder area is given by

$$F_1 = p \times d \times 1$$

this force is resisted by a uniform tangential stress σ_t that is distributed over the cut faces of the cylinder wall. The force generated by this stress is given by

$$F_2 = \sigma_t \times 2t \times 1$$

For equilibrium

$$F_1 = F_2$$

thus

$$\sigma_t = \frac{pd}{2t} \qquad (2.5)$$

Consider now the part longitudinal section of the cylinder including a closed end which is shown in figure 2.5b. The section is formed in a similar way as before. The longitudinal force due to the pressure acting on half the closed end is balanced by the force generated by the longitudinal stress σ_L distributed over the cut face of the cylinder wall, thus for the section shown

$$\frac{p}{2} \left(\frac{\pi d^2}{4} \right) = \frac{\pi d}{2} \sigma_L t$$

or

$$\sigma_L = \frac{pd}{4t} = \frac{\sigma_t}{2} \qquad (2.6)$$

An element of the cylinder wall that has edges respectively parallel and at right angles to the cylinder axis is therefore subject to tensile stresses σ_t and σ_L as shown in figure 2.6.

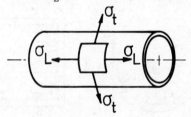

Figure 2.6

The longitudinal deformation of the element consists of an elongation due to σ_L and a contraction due to σ_t because of the Poisson effect. Thus the longitudinal strain, ε_L, of the cylinder is given by

$$\varepsilon_L = \frac{\sigma_L}{E} - \frac{\nu \sigma_t}{E}$$

or, from equations 2.5 and 2.6

$$\varepsilon_L = \frac{pd}{4tE} (1 - 2\nu) \qquad (2.7)$$

Similarly the tangential or circumferential strain, ε_t, is given by

$$\varepsilon_t = \frac{\sigma_t}{E} - \frac{\nu\sigma_L}{E}$$

and from equations 2.5 and 2.6

$$\varepsilon_t = \frac{pd}{4tE} (2 - \nu) \tag{2.8}$$

Since the circumference of the cylinder is simply the diameter multiplied by π, it will be seen that the circumferential strain is equal to the diametral strain.

Let us now determine the change in volume of a thin cylinder of length L under internal pressure. If the change in length and diameter are ΔL and Δd respectively we have

$$L + \Delta L = L(1 + \varepsilon_L)$$

and $d + \Delta d = d(1 + \varepsilon_t)$

Let the original volume of the cylinder be V and the change in volume be ΔV, thus

$$V + \Delta V = (L + \Delta L) \frac{\pi}{4} (d + \Delta d)^2$$

or $V + \Delta V = V(1 + \varepsilon_L)(1 + \varepsilon_t)^2$

then ignoring products of strains we have

$$\Delta V = V(\varepsilon_L + 2\varepsilon_t)$$

after substituting for ε_L and ε_t from equations 2.7 and 2.8 we have

$$\Delta V = V \frac{pd}{4tE} (5 - 4\nu) \tag{2.9}$$

but $V = \frac{\pi d^2 L}{4}$

therefore

$$\Delta V = \frac{\pi d^3 L p}{16E} (5 - 4\nu)$$

alternatively the volumetric strain is given by

$$\frac{\Delta V}{V} = \frac{pd}{4tE} (5 - 4\nu)$$

Example 2.5

A thin cylinder with closed ends has a diameter of 50 times the wall thickness. The cylinder contains air at a pressure of $0 \cdot 7$ MN m^{-2} above the external pressure. The longitudinal and tangential strains due to this pressure are measured and found to be $+ 37 \cdot 2 \times 10^{-6}$ and $145 \cdot 2 \times 10^{-6}$ respectively. Use this information to determine Poisson's ratio and Young's modulus for the cylinder material.

From equations 2.7 and 2.8 and noting that $d = 50t$ we have

$$(0 \cdot 7)50(1 - 2\nu) = 4E(37 \cdot 2 \times 10^{-6}) \tag{1}$$

and $(0 \cdot 7)50(2 - \nu) = 4E(145 \cdot 2 \times 10^{-6})$ (2)

dividing equation 2 by equation 1 we obtain

$$\frac{(2 - \nu)}{(1 - 2\nu)} = \frac{(145 \cdot 2)}{(37 \cdot 2)} = 3 \cdot 9$$

hence $\nu = 0 \cdot 28$

Substituting for ν in either equation 1 or 2 we find that

$$E = 104 \times 10^3 \text{ MN m}^{-2}$$

Example 2.6

A steel cylinder 20 m long having a mean diameter of 3 m and a wall thickness of 20 mm is to contain compressed air at a pressure of 4 MN m^{-2}. Determine the longitudinal and tangential stresses and the changes in length and diameter caused by the internal pressure. What axial force would prevent the longitudinal expansion of the cylinder? Poisson's ratio is $0 \cdot 3$ and Young's modulus is 210 GN m^{-2}.

From equations 2.5 and 2.6 we have

$$\sigma_t = \frac{(4)(3000)}{2(20)} = 300 \text{ MN m}^{-2}$$

and $\sigma_L = \frac{\sigma_t}{2} = 150$ MN m^{-2}

From equations 2.7 and 2.8

$$\varepsilon_t = \frac{(4)(3000)(2 - 0 \cdot 3)}{4\ (20)(210 \times 10^3)} = 1 \cdot 2 \times 10^{-3}$$

thus $\Delta d = 1 \cdot 2 \times 10^{-3} \ (3000) = 3 \cdot 6$ mm

also $\varepsilon_L = \frac{(4)(3000)(1 - 0 \cdot 6)}{4(20)(210 \times 10^3)} = 0 \cdot 28 \times 10^{-3}$

then $\Delta L = 0 \cdot 28 \times 10^{-3} \ (20000) = 5 \cdot 7$ mm

The longitudinal strain, ε'_L, in the cylinder caused by a compressive axial force F acting alone is given by

$$\varepsilon'_L = - \frac{F}{\pi dtE}$$

We require the total longitudinal strain to be zero when F and the internal pressure act simultaneously thus

$$\varepsilon_L + \varepsilon'_L = 0$$

or $0 \cdot 28 \times 10^{-3} - \dfrac{F}{\pi (3)(0 \cdot 02)(210 \times 10^3)} = 0$

hence

$$F = 11 \cdot 3 \text{ MN}$$

2.4   THIN SPHERES

If a thin hollow sphere of mean diameter d and wall thickness t under internal pressure p is cut by a plane passing through its centre, we require for equilibrium that

$$p \frac{\pi d^2}{4} = \sigma_t \pi d t$$

or $\sigma_t = \dfrac{pd}{4t}$ (2.10)

where σ_t is the tangential stress that is distributed over the cut face of the wall of the sphere.

Since the cutting plane referred to above may be chosen quite arbitrarily, the same stress will act in all directions tangential to the surface. Thus the stress in the wall on planes at right angles is σ_t and the tangential strain, which we have seen is equal to the diametral strain, is given by

$$\varepsilon_t = \frac{\sigma_t}{E} - \nu \frac{\sigma_t}{E}$$

or $\varepsilon_t = \dfrac{pd}{4tE}(1 - \nu)$ (2.11)

The change in volume, ΔV, of the sphere under a pressure p is given by

$$V + \Delta V = \frac{\pi}{6}(d + \Delta d)^3$$

where V, the original volume, is given by

$$V = \frac{\pi d^3}{6}$$

but $d + \Delta d = d(1 + \varepsilon_t)$

thus if products of strains are ignored, we have

$$\Delta V = \frac{\pi d^4 p}{8tE}(1 - \nu)$$ (2.12)

alternatively the volumetric strain

$$\frac{\Delta V}{V} = \frac{3pd}{4tE}(1 - \nu)$$ (2.13)

Example 2.7

A spherical steel pressure-vessel has a mean diameter of 400 mm and a wall thickness of 5 mm. Determine the increase in volume and the change in wall thickness under a pressure of 8 MN m^{-2}. E = 210 GN m^{-2} and ν = 0·3.

From equation 2.12

$$\Delta V = \frac{\pi (0·4)^4 (8)(1 - 0·3)}{8(0·005)(210 \times 10^3)} \text{ m}^3$$

$$= 53·6 \times 10^{-6} \text{ m}^3$$

The radial strain, ε_r, in the wall is provided by the tangential tensile stresses, σ_t, acting on planes at right angles since the radial strain due to the internal pressure is negligible, thus

$$\varepsilon_r = -\frac{\nu \sigma t}{E} - \frac{\nu \sigma t}{E} = -\frac{2\nu \sigma t}{E}$$

The new wall thickness is therefore given by

$$t + \Delta t = t(1 + \varepsilon_r) = t \left(1 - \frac{2\nu \sigma t}{E} \right)$$

hence the change in wall thickness

$$\Delta t = -\frac{\nu p d}{2E}$$

$$= -\frac{(0·3)(8)(0·4)}{2(210 \times 10^3)} \text{ m}$$

$$= -2·3 \times 10^{-3} \text{ mm}$$

Example 2.8

A cylindrical steel pressure-vessel having a wall thickness of 5 mm is closed with hemispherical steel ends. Determine the wall thickness of the ends that will ensure that no bending will be induced at the joint. ν = 0·3.

Let the wall thicknesses of the hemispherical end and the cylinder be t_s and t_c respectively. The free diametral strains at the joint for an internal pressure p will be

$$\varepsilon_c = \frac{pd}{4t_c E} (2 - \nu) \text{ for the cylinder}$$

and $\quad \varepsilon_s = \dfrac{pd}{4t_s E} (1 - \nu) \text{ for the hemisphere}$

If these strains are unequal, radial forces distributed round the joint will be necessary to bring the two edges into alignment. These radial forces cause bending. To satisfy the requirements of no bending we must therefore have that

54

$$\varepsilon_c = \varepsilon_s$$

or $\quad \dfrac{t_s}{t_c} = \dfrac{1 - \nu}{2 - \nu} = \dfrac{0 \cdot 7}{1 \cdot 7}$

thus $t_s = 0 \cdot 41 t_c$

hence

$$t_s = 0 \cdot 41 \ (5) = 2 \cdot 06 \ \text{mm}$$

2.6 BULK MODULUS

When liquids are held under pressure, a change in their volume will occur due to compressibility. There is a linear relationship between volumetric strain and pressure valid for small strains such that

$$p = K \frac{\Delta V}{V} \tag{2.14}$$

where K is the bulk modulus of the liquid.

Solids also possess the property of compressibility defined by equation 2.14. The compressibility of gases on the other hand is governed by more complicated relationships belonging to the field of thermodynamics. For the purposes of pressure-vessel design the change in volume of the container will not usually result in a significant change in gas pressure.

Example 2.9

A spherical steel pressure-vessel of 4 m diameter and 100 mm wall thickness is initially full of water at zero gauge-pressure. How much more water must be pumped in to raise the pressure by 5 MN m^{-2}? For steel, $\nu = 0 \cdot 3$ and $E = 210$ GN m^{-2}, K for water $= 2 \cdot 0$ GN m^{-2}.

Under pressure p, the increase in volume of the sphere is given by equation 2.13, thus

$$\Delta V_s = \frac{3pdV}{4tE} \ (1 - \nu) \tag{1}$$

The decrease in volume of the water is given by equation 2.14, thus

$$\Delta V_w = \frac{pV}{K} \tag{2}$$

The extra volume, v, of water at pressure p that must be pumped into the pressure vessel is therefore

$$v = \Delta V_s + \Delta V_w$$

or $v = pV\left[\dfrac{1}{K} + \dfrac{3d}{4tE}(1 - \nu)\right]$

Substituting numerical values we have

$$v = \frac{\pi(5)}{6}(4^3)\left[\frac{1}{2\times 10^3} + \frac{3(4)(0\cdot 7)}{4(0\cdot 1)(210\times 10^3)}\right] \text{ m}^3$$

$$= 100\cdot 53 \times 10^{-3}\text{ m}$$

The volume at atmospheric pressure is given by

$$v' = v\left[1 + \frac{p}{K}\right]$$

$$= 100\cdot 53\left[1 + \frac{5}{2\times 10^3}\right] \times 10^{-3}\text{ m}^3$$

$$= 100\cdot 8 \times 10^{-3}\text{ m}^3$$

The effect of the compressibility of the water is significant since if this had been neglected we should have obtained

$$v' = 16\cdot 75 \times 10^{-3}\text{ m}^3$$

2.5.1 A Relationship between E, K and ν

For a linearly elastic solid it is possible to determine a relationship between the bulk modulus, Young's modulus and Poisson's ratio.

Consider a cube of side a under uniform pressure p. The strain, ε, in each of the directions normal to the cube faces is given by

$$\varepsilon = -\frac{p}{E} + \frac{\nu p}{E} + \frac{\nu p}{E} = -\frac{p}{E}(1 - 2\nu)$$

The new volume of the cube is thus

$$V + \Delta V = a^3(1 + \varepsilon)^3$$

The change in volume if products of strains are ignored is

$$\Delta V = 3\varepsilon a^3 = 3\varepsilon V$$

Thus the volumetric strain is three times the linear strain or

$$\frac{\Delta V}{V} = 3\varepsilon = -\frac{3p}{E}(1 - 2\nu)$$

but from equation 2.14 we see that the volumetric strain is also equal to the pressure divided by the bulk modulus, thus

$$-\frac{p}{K} = -\frac{3p}{E}(1 - 2\nu)$$

or $E = 3K(1 - 2\nu)$ (2.15)

For steel, E = 210 GN m^{-2}, ν = 0·3, thus

$$K = \frac{(210)}{3(1 - 0·6)} = 175 \text{ GN m}^{-2}$$

Since the volumetric strain of a body under pressure can never be positive we have a theoretical upper limit for ν of 0·5.

2.7 STATICALLY INDETERMINATE SYSTEMS

In chapter 1 the idea of statical indeterminacy was introduced. It was pointed out that certain problems could not be solved solely by the application of the equations of statical equilibrium. These statically indeterminate problems require the consideration of deformations for their solution.

In order to derive equations involving deformations it is necessary to make a statement in mathematical terms, of the way in which the different parts of the structure fit together. This statement defines the *compatibility of strains*.

Statically indeterminate problems require the consideration of forces and deformations in the structure. These two factors are related by the *stress-strain characteristics* of the materials of which the structure is composed.

The three basic concepts essential for the solution of statically indeterminate problems in solid mechanics are therefore the satisfaction of the requirements of equilibrium and compatibility linked by a knowledge of the material characteristics.

The following examples illustrate the application of these three concepts.

Example 2.10

A horizontal lever 3 m long is pivoted at one end and carries a load of 5 kN at the other. The lever is supported by two vertical rods of

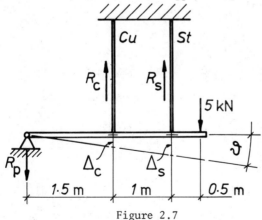

Figure 2.7

equal length each having a cross-sectional area of 60 mm^2. One rod is copper and is pinned at a point 1·5 m from the hinge, the other is steel which is pinned at 2·5 m from the hinge. Assuming the lever to be rigid, determine the stress in each rod. E for steel = 210 GN m^{-2}, E for copper = 105 GN m^{-2}.

Two equations of equilibrium may be obtained by summing the vertical forces and by taking moments about the hinge, thus referring to figure 2.7

$$R_p + 5 = R_c + R_s \text{ kN} \tag{1}$$

and

$$1 \cdot 5R_c + 2 \cdot 5R_s = 3(5) \text{ kN m} \tag{2}$$

We have three unknown forces and two equilibrium equations so the problem is clearly statically indeterminate. To satisfy the requirement of compatibility we see that the angle θ turned through by the rigid lever is given by

$$\theta = \frac{\Delta_c}{1 \cdot 5} = \frac{\Delta_s}{2 \cdot 5} \tag{3}$$

where Δ_c and Δ_s are the deflexions of the copper and steel rods respectively. If the original length of the rods is L, equation 3 in terms of strains becomes

$$\frac{\varepsilon_c L}{1 \cdot 5} = \frac{\varepsilon_s L}{2 \cdot 5}$$

or

$$\varepsilon_c = \frac{3\varepsilon_s}{5} \tag{4}$$

where ε_c and ε_s are the strains in the rods.

The axial forces R_c and R_s in the rods produce stresses σ_c and σ_s. These stresses are related to the strains by the elastic characteristics

$$\sigma_c = E_c \varepsilon_c \text{ and } \sigma_s = E_s \varepsilon_s$$

thus $\varepsilon_c = \dfrac{R_c}{E_c A}$ and $\varepsilon_s = \dfrac{R_s}{E_s A}$

where A is the cross-sectional area of the rods.

Substituting for the strains in equation 4 we have

$$\frac{R_s}{E_c A} = \frac{3}{5} \times \frac{R_s}{E_s A}$$

or

$$R_c = \frac{3}{5}\left(\frac{105}{210}\right) R_s$$

thus

$$R_c = 0 \cdot 3R_s \tag{5}$$

Substituting for R_c in equation 2 we obtain

58

$$1 \cdot 5(0 \cdot 3R_s) + 2 \cdot 5R_s = 15 \text{ kN m}$$

hence

$$R_s = 5 \cdot 085 \text{ kN}$$

then from equation 5

$$R_c = 1 \cdot 525 \text{ kN}$$

The cross-sectional area of both rods is 60 mm² thus

$$\sigma_c = 25 \cdot 4 \text{ N mm}^{-2}$$

and $\sigma_s = 84 \cdot 7 \text{ N mm}^{-2}$

From equation 1 the vertical reaction at the pivot is

$$R_p = 5 \cdot 085 + 1 \cdot 525 - 5 = 1 \cdot 61 \text{ kN}$$

Example 2.11

A steel tube of length 2 m, mean diameter 0·5 m and wall thickness 10 mm is filled with concrete. The resulting column is subjected to an axial compressive force of 4 MN through rigid end-plates. Determine the proportions of the load that are taken by the steel and the concrete and the contraction of the column. Ignore the effect of tangential stresses in the tube. E for concrete = 14 GN m^{-2}, E for steel = 210 GN m^{-2}.

Let the force in the tube be P_S and the force in the concrete be P_C then

$$P_S + P_C = 4 \text{ MN} \tag{1}$$

If the contraction in the column length is Δ mm, the common strain, ε, is given by

$$\varepsilon = \frac{\Delta}{2000} \tag{2}$$

but $\varepsilon = \dfrac{P_S}{E_S A_S} = \dfrac{P_C}{E_C A_C}$ $\tag{3}$

where A_S and A_C are the cross-sectional areas of the steel and concrete.

From equation 3

$$P_S = \left(\frac{210}{14}\right)\frac{\pi(0 \cdot 5)(0 \cdot 01)}{\pi(0 \cdot 5)^2} 4P_C$$

or $P_S = 1 \cdot 2P_C$

Substituting for P_S in equation 1 we have

$$P_c = \frac{4}{2 \cdot 2} = 1 \cdot 82 \text{ MN}$$

thus $P_s = 2 \cdot 18$ MN

From equations 2 and 3

$$\Delta = \frac{2000(2 \cdot 18)}{(210 \times 10^3)\pi(0 \cdot 5)(0 \cdot 01)} = 1 \cdot 32 \text{ mm}$$

Example 2.12

A cylindrical pressure-vessel of mean diameter 50 mm and wall thickness 6 mm has its ends welded to flat circular plates. The end plates are tied together by four rods of diameter 5 mm arranged symmetrically around the outside of the cylinder. Determine the stresses in the cylinder and tie-rods due to an internal fluid pressure of 6 MN m^{-2}. The cylinder and rod are of steel for which $\nu = 0 \cdot 3$. Assume that the cylinder and rods are the same length.

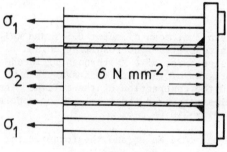

Figure 2.8

Referring to figure 2.8, we have, for equilibrium of horizontal forces

$$\sigma_1 \times 4 \left(\frac{\pi 5^2}{4}\right) + \sigma_2 \ (\pi \times 50 \times 6) = 6\left(\frac{\pi 50^2}{4}\right)$$

hence

$$\sigma_1 + 12\sigma_2 = 150 \text{ N mm}^{-2} \tag{1}$$

Since strains in the rod and cylinder are equal we have

$$\frac{\sigma_1}{E} = \frac{\sigma_2}{E} - \frac{\nu\sigma_t}{E}$$

where σ_t is the tangential stress in the cylinder, then

$$\sigma_1 = \sigma_2 - \frac{(0 \cdot 3)(6)(50)}{2(6)}$$

or $\quad \sigma_1 - \sigma_2 = - 7 \cdot 5 \text{ N mm}^{-2} \tag{2}$

From equations 1 and 2

$$\sigma_1 = 4 \cdot 6 \text{ N mm}^{-2}$$

and $\sigma_2 = 12 \cdot 1 \text{ N mm}^{-2}$

2.8 THERMAL EFFECTS

When materials are subjected to temperature changes they respond by expanding or contracting. These dimensional changes give rise to thermal strains that do not produce stresses in the materials unless the dimensional change is partially or wholly prevented.

Suppose we consider the change in length of a bar (initial length L) caused by a rise in temperature T K. If the bar is free to expand, the change in length, Δ, is given by

$$\Delta = \alpha L T \qquad\qquad\qquad\qquad (a)$$

where α is the coefficient of linear expansion for the bar material. We shall assume here that the coefficient is a constant.

Since the bar is free to expand no stresses will be induced. However, suppose we prevent the bar expanding; to do this we must compress the bar with a force P to restore it to its original length. If the stress due to P is σ then the deformation Δ is given by

$$\Delta = \frac{\sigma}{E} (L + \Delta) \qquad\qquad\qquad\qquad (b)$$

hence

$$\sigma = \frac{E\alpha T}{1 + \alpha T}$$

but αT is very small compared with unity, thus we may write

$$\sigma = E\alpha T \qquad\qquad\qquad\qquad (c)$$

If materials having different coefficients of thermal expansion are combined in a structural system, internal forces may be generated

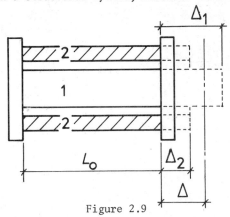

Figure 2.9

when temperature changes occur. Consider the problem of a bar of material 1 inside a tube of material 2 to which it is rigidly attached at each end. We require the stresses induced by a temperature rise of T K.

Figure 2.9 shows the free thermal expansions that would occur if the end plate were not present. If material 1 has a greater coefficient of linear expansion than material 2 we have

$$\Delta_1 = L_0\alpha_1 T \text{ and } \Delta_2 = L_0\alpha_2 T$$

The actual extensions of the bar and the tube must be the same and to achieve this the bar must be compressed and the tube stretched by a pair of equal and opposite internal forces. If the common extension of the bar and tube is Δ, the strains produced by the internal forces are

$$\varepsilon_1 = \frac{\Delta - \Delta_1}{L_0} \text{ and } \varepsilon_2 = \frac{\Delta - \Delta_2}{L_0}$$

if Δ_1 and Δ_2 are small compared with L_0.

The corresponding stresses are

$$\sigma_1 = E_1 \frac{(\Delta - \Delta_1)}{L_0} \tag{a}$$

and $$\sigma_2 = E_2 \frac{(\Delta - \Delta_2)}{L_0} \tag{b}$$

For equilibrium of the internal forces we have

$$\sigma_1 A_1 + \sigma_2 A_2 = 0 \tag{c}$$

From equations a, b and c and the expressions for Δ_1 and Δ_2 we obtain

$$\frac{E_1 A_1}{L_0} (\Delta - L_0\alpha_1 T) + \frac{E_2 A_2}{L_0} (\Delta - L_0\alpha_2 T) = 0$$

hence

$$\Delta = \frac{(E_1 A_1 \alpha_1 + E_2 A_2 \alpha_2)}{(E_1 A_1 + E_2 A_2)} L_0 T$$

Substituting for Δ in equations a and b we obtain the stresses,

thus $$\sigma_1 = \frac{E_1 E_2 A_2 (\alpha_2 - \alpha_1) T}{E_1 A_1 + E_2 A_2}$$

and $$\sigma_2 = \frac{E_1 E_2 A_1 (\alpha_1 - \alpha_2) T}{E_1 A_1 + E_2 A_2}$$

Example 2.13

A bimetallic rod of length 450 mm is mounted horizontally between rigid abutments. The rod has a uniform circular cross-section and is made up of a 150 mm length of steel and a 300 mm length of copper coaxial with each other. If the rod is initially stress-free, determine the stress in the rod caused by a temperature rise of 100 K.

E (copper) = 105 GN m^{-2}
E (steel) = 210 GN m^{-2}
α (copper) = 18 × 10^{-6} K^{-1}
α (steel) = 12 × 10^{-6} K^{-1}

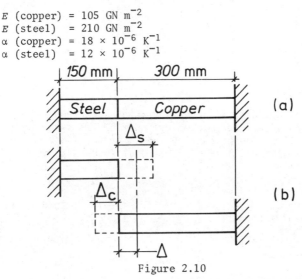

Figure 2.10

Figure 2.10a shows the arrangement of the bars and figure 2.10b the free thermal displacements Δ_s and Δ_c together with the final common displacement, Δ, of the interface.

We have

$$\Delta_s = (150)(100)(12 \times 10^{-6}) = 0\cdot18 \text{ mm}$$

and $\Delta_c = (300)(100)(18 \times 10^{-6}) = 0\cdot54$ mm

Suppose the final length of the steel bar is $(150 + \Delta)$. The final length of the copper bar is therefore $(300 - \Delta)$. The strains in the two bars are thus

$$\varepsilon_s = \frac{(150 + \Delta_s) - (150 + \Delta)}{150} = \frac{\Delta_s - \Delta}{150} \tag{1}$$

and $\varepsilon_c = \dfrac{(300 + \Delta_c) - (300 - \Delta)}{300} = \dfrac{\Delta_c + \Delta}{300}$ $\qquad$ (2)

It is assumed here that the free thermal expansions are small compared with the original lengths of the bars.

If the force exerted at the interface produces a stress σ in the bars then

$$\sigma = E_s \epsilon_s = E_c \epsilon_c \tag{3}$$

and from equations 1 and 2 we have

$$E_s \frac{(\Delta_s - \Delta)}{150} = E_c \frac{(\Delta_c + \Delta)}{300}$$

hence $\Delta = 0 \cdot 036$ mm

From equation 3

$$\sigma = \frac{E_s(\Delta_s - \Delta)}{150}$$

or $\quad \sigma = 201 \cdot 6$ MN m^{-2}

Example 2.14

A hollow steel cylinder of cross-sectional area 2000 mm^2 concentrically surrounds a solid aluminium cylinder of cross-sectional area 6000 mm^2. Both cylinders have the same length of 500 mm before a rigid block weighing 200 kN is applied at 20°C as shown in figure 2.11a. Determine

(a) the load carried by each cylinder at 60°C
(b) the temperature rise required for the total load to be carried by the aluminium

E (steel) = 210 GN m^{-2}
E (aluminium) = 70 GN m^{-2}
α (steel) = 12 × 10^{-6} K^{-1}
α (aluminium) = 23 × 10^{-6} K^{-1}

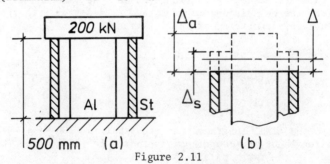

Figure 2.11

Let subscripts a and s stand for aluminium and steel respectively.

Figure 2.11b shows the free thermal expansions Δ_a and Δ_s together with the common expansion Δ under the load of 200 kN.

For a temperature rise of T K we have

$$\Delta_a = 500 \times 23T \times 10^{-6} = 11 \cdot 5T \times 10^{-3} \text{ mm}$$

and $\quad \Delta_s = 500 \times 12T \times 10^{-6} = 6T \times 10^{-3}$ mm

64

Under load, the strains are therefore

$$\epsilon_a = \frac{\Delta_a - \Delta}{500} \text{ and } \epsilon_s = \frac{\Delta_s - \Delta}{500}$$

and the corresponding stresses are

$$\sigma_a = \frac{70 \times 10^3}{500} (\Delta_a - \Delta) = 140 (\Delta_a - \Delta) \text{ N mm}^{-2}$$

and $\sigma_s = \frac{210 \times 10^3}{500} (\Delta_s - \Delta) = 420 (\Delta_s - \Delta) \text{ N mm}^{-2}$

For equilibrium of vertical forces

$$\sigma_a \times 6000 + \sigma_s \times 2000 = 200 \times 10^3 \text{ N}$$

Substituting for σ_a, σ_s, Δ_a and Δ_s we have

$$(11 \cdot 5T \times 10^{-3} - \Delta) + (6T \times 10^{-3} - \Delta) = \frac{5}{21}$$

hence

$$\Delta = 8 \cdot 75T \times 10^{-3} - \frac{5}{42}$$

The loads taken by the aluminium and the steel are therefore

$$P_a = \sigma_a \times 6000 \text{ N}$$

$$= 840 \left(2 \cdot 75T \times 10^{-3} + \frac{5}{42} \right) \text{ kN}$$

and $P_s = \sigma_s \times 2000 \text{ N}$

$$= 840 \left(\frac{5}{42} - 2 \cdot 75T \times 10^{-3} \right) \text{ kN}$$

These equations are valid provided Δ does not exceed Δ_s. All the load will be taken by the aluminium when $\Delta_s = \Delta$, thus the temperature rise is given by

$$6T \times 10^{-3} = 8 \cdot 75T \times 10^{-3} - \frac{5}{42}$$

or $T = \frac{5 \times 10^3}{2 \cdot 75 \times 42} = 43 \cdot 3 \text{ K}$

The loads carried by the aluminium and the steel at a temperature of 60°C ($T = 40$ K) are

$$P_a = 840 \left(2 \cdot 75 \times 40 \times 10^{-3} + \frac{5}{42} \right) \text{ kN}$$

$$= 192 \cdot 4 \text{ kN}$$

and $P_s = 200 - 192 \cdot 4 = 7 \cdot 6 \text{ kN}$

Example 2.15

A steel bolt of diameter 12 mm and length 175 mm is used to clamp a brass sleeve of length 150 mm to a rigid base plate as shown in figure 2.12. The sleeve has an internal diameter of 25 mm and a wall thickness of 3 mm. The thickness of the base plate is 25 mm. Initially the nut is tightened until there is a tensile force of 5 kN in the bolt. The temperature is now increased by 100 K. Determine the final stresses in the bolt and the sleeve.

E (brass) = 105 GN m^{-2}
E (steel) = 210 GN m^{-2}
α (brass) = 20 × 10^{-6} K^{-1}
α (steel) = 12 × 10^{-6} K^{-1}

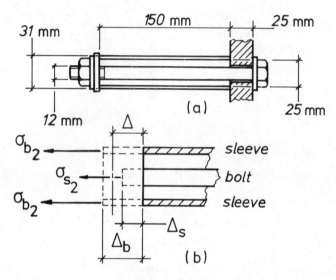

Figure 2.12

Figure 2.12a shows the dimensions and general arrangement of the assembly. The initial stresses in the steel and brass due to the 5 kN load are

$$\sigma_{s_1} = + \frac{5 \times 10^3 \times 4}{\pi (12^2)} = + 44 \cdot 21 \text{ N mm}^{-2}$$

and $$\sigma_{b_1} = - \frac{5 \times 10^3 \times 4}{\pi (31^2 - 25^2)} = - 18 \cdot 95 \text{ N mm}^{-2}$$

where subscripts s and b stand for steel and brass respectively.

The free thermal expansions Δ_s and Δ_b of the steel and brass are shown in figure 2.12b together with their common expansion Δ.

$$\Delta_s = (175)(12 \times 10^{-6})(100) = 0 \cdot 21 \text{ mm}$$

and $$\Delta_b = (150)(20 \times 10^{-6})(100) = 0 \cdot 3 \text{ mm}$$

66

Equilibrium of the thermal stresses σ_{s_2} and σ_{b_2} requires that

$$\sigma_{s_2} \frac{\pi(12^2)}{4} + \sigma_{b_2} \frac{\pi(31^2-25^2)}{4} = 0$$

therefore

$$3\sigma_{s_2} + 7\sigma_{b_2} = 0 \qquad\qquad (1)$$

The thermal strains are given by

$$\varepsilon_s = \frac{(\Delta - \Delta_s)}{175} \text{ and } \varepsilon_b = \frac{(\Delta - \Delta_b)}{150}$$

thus the thermal stresses are

$$\sigma_{s_2} = \frac{210}{175} (\Delta - \Delta_s)\ 10^3 \text{ N mm}^{-2}$$

and $\sigma_{b_2} = \dfrac{105}{150} (\Delta - \Delta_b)\ 10^3 \text{ N mm}^{-2}$

Substituting in equation 1

$$3 \frac{210}{175} (\Delta - 0 \cdot 21)\ 10^3 + 7 \frac{105}{150} (\Delta - 0 \cdot 30)\ 10^3 = 0$$

hence

$$\Delta = 0 \cdot 262 \text{ mm}$$

thus $\sigma_{s_2} = \dfrac{210}{175} (0 \cdot 262 - 0 \cdot 21)\ 10^3 = + 62 \cdot 26 \text{ N mm}^{-2}$

and $\sigma_{b_2} = \dfrac{105}{150} (0 \cdot 262 - 0 \cdot 30)\ 10^3 = - 26 \cdot 68 \text{ N mm}^{-2}$

Total stresses are therefore

$$\sigma_s = \sigma_{s_1} + \sigma_{s_2} = + 106 \cdot 5 \text{ N mm}^{-2}$$

and $\sigma_b = \sigma_{b_1} + \sigma_{b_2} = - 45 \cdot 6 \text{ N mm}^{-2}$

Example 2.16

A spherical vessel of internal diameter d and wall thickness t is completely filled with water and sealed. Obtain an expression for the increase in pressure due to a rise in temperature of the system.

The original internal volume of the vessel, V_0 is given by

$$V_0 = \frac{\pi d^3}{6}$$

The total tangential strain in the wall of the vessel is made up of the strain due to internal pressure p and the thermal strain due to a temperature rise T, thus

$$\varepsilon_T = \alpha T + \frac{pd}{4tE} (1 - \nu)$$

The new volume of the vessel V_s, is thus

$$V_s = V_0 (1 + \varepsilon_T)^3$$

since the tangential strain is equal to the diametral strain. If we ignore products of strains we have

$$V_s = \frac{\pi d^3}{6} \left(1 + 3\alpha T + \frac{3pd}{4tE}(1 - \nu) \right) \tag{1}$$

The volume of the water will increase due to the rise in temperature and decrease due to the increase in pressure. If the bulk modulus of water is K, the volumetric strain due to the pressure is

$$\frac{dV_1}{V_0} = - \frac{p}{K}$$

Suppose that the coefficient of volumetric expansion for the water is β, then the volumetric strain due to the temperature rise is

$$\frac{dV_2}{V_0} = \beta T$$

The new volume of the water is thus

$$V_w = \frac{\pi d^3}{6} \left(1 - \frac{p}{K} + \beta T \right) \tag{2}$$

The two volumes given by equations 1 and 2 must be equal, thus

$$3\alpha T + \frac{3pd}{4tE} (1 - \nu) = \beta T - \frac{p}{K}$$

hence

$$p = \frac{4tEK \ (\beta - 3\alpha) \ T}{4tE + 3Kd \ (1 - \nu)}$$

2.9 PROBLEMS FOR SOLUTION

1. A composite shaft consists of a brass bar of 50 mm diameter and 200 mm length to each end of which are concentrically friction-welded steel rods of 20 mm diameter and 100 mm length. At a particular stage during a tensile test on the composite bar, the overall extension is measured as 0·15 mm. What are then the axial stresses in the two parts of the bar?

E (brass) = 120 GN m^{-2}
E (steel) = 210 GN m^{-2}

(123 N mm^{-2} in steel, 19·7 N mm^{-2} in brass)

2. A steel bar of rectangular section 30 mm × 6 mm is subjected to an

axial pull of 35 kN when it is found that the extension measured on a length of bar of 200 mm is 0·19 mm.

The steel bar is then sandwiched between two aluminium alloy bars each of rectangular section 30 mm × 8 mm to form a composite bar of rectangular section 30 mm wide and 22 mm thick. An axial pull of 70 kN is applied to the composite bar when it is found that the extension measured on a length of 200 mm is 0·21 mm.

Determine the values of Young's modulus for the steel and the alloy.
(204·7 and 62·1 GN m^{-2})

3. Two vertical rods are each rigidly fastened at the upper end at a distance apart of 600 mm. Each rod is 3 m long and 12 mm in diameter. A horizontal cross-bar connects the lower ends of the rods and on it is placed a load of 5 kN so that the cross-bar remains horizontal. Find the position of the load on the cross-bar and calculate the stress in each rod. One rod is steel, E = 210 GN m^{-1}, and the other is bronze, E = 70 GN m^{-2}.
(150 mm from steel rod; 33 N mm^{-2} in steel, 11 N mm^{-2} in bronze)

4. A small square portion of flat steel plate has tensile strains of $2·37 \times 10^{-4}$ and $2·95 \times 10^{-4}$ in the two directions perpendicular to the sides of the square. Determine the corresponding stresses. E = 210 GN m^{-2} and ν = 0·28.
(72·8 and 82·3 N mm^{-2})

5. A block of material in the form of a cube of unit side is heated to a temperature T from zero. In the x-direction the block is unrestrained, in the y-direction it is completely restrained and in the z-direction it is restrained by another material whose resistance is given by S, stress per unit strain.

Using E for Young's modulus, α for the coefficient of linear expansion and taking Poisson's ratio as 0·3, determine the stresses in the coordinate directions for the case when S = 0·3E. What is the change in volume of the cube?
(0, - 1·092 αET, - 0·306 $E\alpha T$, 2·44αT)

6. Two thin cylinders of equal length and the same material are placed one inside the other. The outer cylinder has twice the diameter and twice the wall thickness of the inner cylinder. The ends of the cylinders are rigidly clamped together. The cylinders are free to expand or contract longitudinally together.

If the inside of the inner cylinder is open to the atmosphere, show that when pressure is applied to the annular space between the cylinders, the ratio of longitudinal stress in the inner cylinder to that in the outer cylinder is given by

$$\frac{3 - 16\nu}{3 + 4\nu}$$

where ν is Poisson's ratio for the material.

7. A straight pipe 3 m long, 60 mm bore and 3 mm wall thickness is blanked off at the ends for hydraulic testing. Determine the volume of water at atmospheric pressure which must be pumped into the pipe to give a tangential stress of 80 MN m^{-2}. Neglect the restraining effect of the end flanges. Take E = 210 GN m^{-2}, ν = 0·28, K = 2·0 GN m^{-2}.
(8·56 litres)

8. In the manufacture of a straight pre-stressed concrete beam of uniform cross-section, the steel bars are subjected to an initial total tensile-load of 2 MN and this load is maintained until the concrete has set. Removal of the load leaves the beam in a pre-stressed state. Determine the residual stresses in the concrete and the steel. The cross-sectional areas of the steel and concrete are 5 × 10^3 and 100 × 10^3 mm^2 respectively. Assume that the centroids of the steel and concrete are coincident and that the ratio of Young's moduli is 15.
(228·6 N mm^{-2} in steel, 11·4 N mm^{-2} in concrete)

9. Figure 2.13 shows a solid, circular-section steel rod supported in a recess and surrounded by a coaxial brass tube. The diameter of the rod is 32 mm. The internal diameter of the tube is 45 mm and its external diameter is 50 mm. Initially the top of the tube is 0·1 mm above the top of the rod. The assembly is now subjected to vertical compression in a testing machine. Determine

  (a) the magnitude of the load if the stresses in the rod and the tube are not to exceed 110 and 80 MN m^{-2} respectively

  (b) the end shortening of the tube when the load is such that the stresses in the tube and rod are equal.

  E (steel) = 210 GN m^{-2}, E (brass) = 105 GN m^{-2}.
(84·1 kN, 0·3 mm)

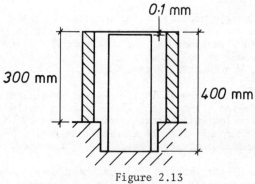

0·1 mm

300 mm

400 mm

Figure 2.13

10. A tube, 100 mm long, 30 mm outside diameter and 25 mm inside diameter, has each end closed by a rigid cover-plate 15 mm thick. The cover plates are secured by means of a rod 20 mm in diameter. The tube and the rod are made of the same material. The rod is central in the tube and the threaded ends of the rod pass freely through holes in the cover plates where they are anchored by nuts.

Initially the nuts are tightened so that the tube just starts to yield at a stress of 250 MN m^{-2}. What is the stress in the rod?

External tensile forces are now applied to the ends of the rod so that the rod just starts to yield. What is now the stress in the tube? Determine the magnitude of the external tensile forces. (171·9, 148·4 N mm^{-2}; 46·5 kN)

11. A straight bar has a circular cross-section the radius of which varies linearly from 30 mm at one end A to 15 mm at the other end B. The bar is 1 m long and is fixed rigidly at A, but longitudinally movement is possible at B against a spring which opposes movement with a constant stiffness of 20 kN mm^{-1}.

Initially there is no longitudinal stress in the bar. The temperature of the bar then falls by 100 K. Determine the change in the bar length if E = 69 GN m^{-2} and α = 23 × 10^{-6} K^{-1}. (-1·26 mm)

12. A thin-walled steel cylinder of mean diameter 70 mm and wall thickness 2 mm is sealed at each end by means of rigid plates pulled together by a central steel rod of diameter 10 mm.

If the unit is assembled with the rod at a temperature of 80K above the cylinder, determine the longitudinal stress in the latter when cooling has occurred. What is the internal pressure necessary to reduce this stress by 80 per cent?

Take E = 200 GN m^{-2}, α = 12 × 10^{-6} K^{-1} and ν = 0·28. (29·1 N mm^{-2}, 2900 kN m^{-2})

3 TORSION

We have seen in the previous chapter that axial loads produce normal
stresses. Torques and shear forces on the other hand induce shear
stresses. Although we shall not be concerned with shear forces in
this chapter, it will be useful to refer to shear force in order to
introduce the concept of shear stress.

3.1 SHEAR STRESS

Unlike normal stresses which are set up on planes normal to the
direction of the applied load, shear stresses appear on planes
parallel to the direction of the load.

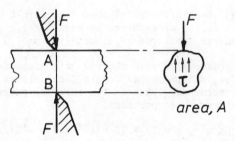

Figure 3.1

Imagine the forces, F, applied to a bar by a pair of shears
(figure 3.1). The forces from the blades tend to separate material
along the line AB. If the lower blade and the bar to the right of
AB are removed, the shear force F from the upper blade is seen to be
resisted by shear stresses τ (tau) distributed over the cross-
sectional area A of the bar. The distribution of shear stress will
not be uniform since this depends on the shape of the bar cross-
section (see chapter 8). Although the calculation of the actual
shear-stress distribution is complicated, it is very easy to determine the mean shear stress τ_m from

$$\tau_m = \frac{F}{A} \tag{3.1}$$

As for normal stress, the units of shear stress depend on the
units chosen for F and A. The usual units for shear stress are
$N\ mm^{-2}$ or $MN\ m^{-2}$.

3.2 COMPLEMENTARY SHEAR STRESS

Let us now look more closely at the stresses in the bar of figure
3.1. If we isolate a small block of material from somewhere along
the line AB, we find that shear stresses τ_1 exist on vertical faces
as we would expect but in addition to these, there must also be

shear stresses τ_2 on the horizontal faces in order to preserve equilibrium.

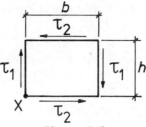

Figure 3.2

Suppose the block is of unit thickness and has a height h and breadth b, (figure 3.2). Clearly horizontal and vertical equilibrium of forces is satisfied. For equilibrium of moments of forces we take moments about point X, thus

$$\tau_1(h \times 1)b = \tau_2(b \times 1)h$$

or $\quad \tau_1 = \tau_2$

Hence the vertical shear stress that arises from the vertical shear force in figure 3.1 must be accompanied by equal complementary shear stress on horizontal planes.

3.3 SHEAR STRAIN

We are now in a position to examine deformations due to shear. The small block ABCD in figure 3.3 is subjected in one plane to vertical shear stresses τ and as we have seen, there must also be equal complementary shear stresses on the horizontal faces.

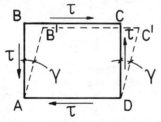

Figure 3.3

The forces produced by the stresses acting on the faces of the block will cause it to distort without change in volume from the original rectangular section ABCD to the parallelogram AB'C'D. The shear strain is defined as the small angle γ through which the vertical faces rotate, thus

$$\gamma = \frac{BB'}{AB} = \frac{CC'}{CD}$$

The shear strain γ is measured in radians and like normal strain is non-dimensional.

For most materials, certainly those used for construction, the following linear relationship between shear stress and shear strain exists, providing the limit of elasticity is not exceeded.

$$\tau = G\gamma \qquad (3.2)$$

The proportionality constant G is called the modulus of rigidity or the shear modulus. It has the units of stress since the shear strain γ is non-dimensional.

3.4 TORSION OF A SOLID CIRCULAR SHAFT

A practical example of a pure shear-stress system occurs with the torsion of a solid shaft of uniform circular cross-section. Consider such a shaft (figure 3.4a) completely restrained at one end by encasement in a wall. The other end of the shaft is loaded in torsion by a pair of equal and opposite forces acting at right angles to a diameter and placed symmetrically on either side of the shaft axis.

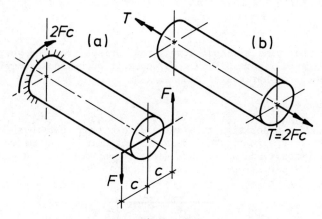

Figure 3.4

The pair of forces constitute a couple of magnitude $2Fc$ which has the sole effect of twisting the shaft about its longitudinal axis. A couple acting in this way is usually referred to as a torque. The couple in figure 3.4a can be replaced by the torque T ($= 2Fc$) shown in figure 3.4b. The torque is indicated vectorially by the double-headed arrow and its direction of rotation is obtained using the right-hand-screw rule. An equal and opposite reactant torque is generated at the wall to preserve equilibrium.

Owing to the symmetry of the shaft we may assume for small deformations that cross-sections will remain plane and radii remain straight during twisting.

To determine the deformation of the twisted shaft, we refer to figure 3.5 which shows a small element of the shaft at a distance z from the wall. The length of the element is δz and the radius of the shaft is r.

74

The torque applied to the whole shaft will result in a system of shear stresses distributed over the circular faces of the element. Let the shear stress at radius r be τ.

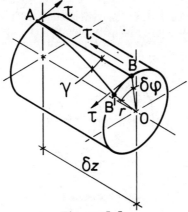

Figure 3.5

Consider now a straight line AB on the surface of the element parallel to the longitudinal axis. Under the action of the shear stresses τ this line will deform into the helix AB'. From our previous definition, we see that for small deformations the angle BAB' is the shear strain γ induced by τ, thus

$$BB' = \gamma AB = \gamma \ \delta z \qquad \qquad (a)$$

Looking now at the circular face of the element, we find that BB' subtends an angle δφ at the axis. This small angle represents the twist of the surface of the element over the length δz, thus

$$BB' = r \ \delta \phi \qquad \qquad (b)$$

From equations a and b we have

$$\gamma \ \delta z = r \ \delta \phi \qquad \qquad (c)$$

Substituting from equation 3.2 for γ we obtain in the limit

$$\tau = Gr \ \frac{d\phi}{dz} \qquad \qquad (d)$$

The total angle of twist φ for a length of shaft L and constant radius r can be determined by integrating equation d. Taking the wall as the datum for φ and the origin of the z-axis, we have

$$\phi = \int_0^\phi d\phi = \int_0^L \frac{\tau}{Gr} \ dz \qquad \qquad (e)$$

Since the torque applied to the shaft is constant over the length, the shear stress τ at radius r must also remain constant so that performing the integration above and rearranging we obtain

$$\tau = Gr \ \frac{\phi}{L} \qquad \qquad (3.3)$$

We see from equation 3.3 that the distribution of shear stress in the cross-section is linear with respect to the radius. It is zero at the axis and a maximum at the surface.

We are now able to relate the magnitude of the shear stress in the shaft to the torque T which produces it. Consider the cross-section of the shaft of radius R shown in figure 3.6.

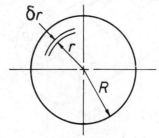

Figure 3.6

An annular element of width δr at radius r will carry a uniform shear stress τ. The total area of the element is $2\pi r\, \delta r$. The force δQ produced by the shear stress acting on the element is therefore given by

$$\delta Q = 2\pi r\, \delta r\, \tau \tag{f}$$

Since this force acts at a constant radius r, it produces a torque δT given by

$$\delta T = r\, \delta Q = 2\pi r^2 \tau\, \delta r \tag{g}$$

Substituting for τ from equation 3.3 we have

$$\delta T = 2\pi\, \frac{G\phi}{L}\, r^3\, \delta r \tag{h}$$

Proceeding to the limit and integrating this expression from $r = 0$ to $r = R$ we obtain the total torque

$$T = \int_0^T \mathrm{d}T = 2\pi\, \frac{G\phi}{L} \int_0^R r^3\, \mathrm{d}r \tag{i}$$

or $\quad T = \dfrac{\pi R^4}{2} \times \dfrac{G\phi}{L}$

The term $\pi R^4/2$ is the polar second moment of area of a solid circular cross-section. It is usually given the symbol J. The final expression for the torque is thus

$$T = GJ\, \frac{\phi}{L} \tag{3.4}$$

Notice that in equation g above we could have substituted for τ from equation d. Since the subsequent integration is carried out over a cross-section for which $\mathrm{d}\phi/\mathrm{d}z$ is a constant, the expression for the torque becomes

$$T = GJ \frac{d\phi}{dz} \tag{3.5}$$

This equation will be of value later when we come to consider the behaviour of tapered shafts or the effect of distributed torque.

Returning to the case of the uniform shaft under a constant torque we have from equations 3.3 and 3.4

$$\frac{T}{J} = \frac{\tau}{r} = \frac{G\phi}{L} \tag{3.6}$$

We shall now illustrate the use of these equations with a numerical example.

Example 3.1

A solid circular-section steel shaft has a radius of 20 mm. The shaft is subjected to a constant torque of 1·2 kN m. Determine the maximum shear stress and the angle of twist over a length of 4 m. The shear modulus G is 80 GN m^{-2}.

The polar second moment of area is given by

$$J = \frac{\pi R^4}{2} = 8\pi \times 10^4 \text{ mm}^4$$

From equations 3.6 the maximum shear stress which occurs on the surface of the shaft is

$$\tau_{max} = \frac{(1 \cdot 2 \times 10^6)(20)}{(8\pi \times 10^4)} \text{ N mm}^{-2} = 95 \cdot 5 \text{ N mm}^{-2}$$

Also from equations 3.6 we have the angle of twist

$$\phi = \frac{TL}{GJ} = \frac{(1 \cdot 2 \times 10^6)(4 \times 10^3)}{(80 \times 10^3)(8\pi \times 10^4)}$$

thus $\phi = 0 \cdot 24$ rad $= 13 \cdot 7°$.

3.5   TORSION OF A HOLLOW CIRCULAR SHAFT

The analysis that led to equation 3.3 for a solid shaft applies equally well to a hollow shaft. There is a change however in the integral expression for the torque, equation j. The range of integration now extends from the internal radius of the shaft R_i to the external radius R_e, thus

$$T = 2\pi \frac{G\phi}{L} \int_{R_i}^{R_e} r^3 \, dr$$

or $\quad T = \frac{\pi}{2} (R_e^4 - R_i^4) \frac{G\phi}{L}$

The term $\pi(R_e^4 - R_i^4)/2$ is the polar second moment of area of a hollow circular cross-section which we again denote by the symbol J. The previous torsion equations (3.6) can therefore be applied to hollow circular sections providing we take the appropriate value for J.

A disadvantage of the solid shaft is that material close to the axis is contributing very little to the torque-carrying capacity. With a hollow shaft, the available material is used more efficiently. Thus for a particular design problem it is often possible to replace a solid shaft with a lighter hollow one carrying the same load.

Example 3.2

It is required to replace the solid shaft of example 3.1 with a hollow shaft of external radius 22 mm. Determine the internal radius for the same torque and maximum shear-stress conditions as the solid shaft and compare their weights and angles of twist.

For the solid shaft

$$\frac{2T}{\pi R^4} = \frac{\tau_{max}}{R} \tag{1}$$

For the hollow shaft

$$\frac{2T}{\pi(R_e^4 - R_i^4)} = \frac{\tau_{max}}{R_e} \tag{2}$$

Since the torque and maximum shear stress are to be the same for both shafts we have from equations 1 and 2 that

$$R_i = R_e \left[1 - \left(\frac{R}{R_e}\right)^3 \right]^{1/4}$$

Now $R = 20$ mm and $R_e = 22$ mm, thus

$$R_i = 22 \left[1 - \left(\frac{20}{22}\right)^3 \right]^{1/4} \text{ mm} = 15 \cdot 5 \text{ mm}$$

The ratio of the weights of the shafts is the same as the ratio of their cross-sectional areas, thus

$$\frac{W_{hollow}}{W_{solid}} = \frac{R_e^2 - R_i^2}{R^2} = 0 \cdot 605$$

Angle of twist is inversely proportional to polar second moment of area, thus

$$\frac{\phi_{hollow}}{\phi_{solid}} = \frac{R^4}{R_e^4 - R_i^4} = 0 \cdot 91$$

The substitution of a hollow shaft has therefore resulted in a weight saving of some 40% and an increased torsional stiffness. These advantages have been paid for by the 10% increase in the external radius of the hollow shaft.

Further improvement is possible if a greater external radius is permitted. For example, if $R_e = 24$ mm, the weight saving becomes about 50% and the angle of twist per unit length falls to 84% of that for the solid shaft. Notice, however, that the minimum external radius for a given torque and maximum shear stress can only be achieved by using a solid shaft.

3.6 POWER TRANSMISSION

The work, W, done by a constant torque T turning through an angle θ is given by

$$W = T\theta$$

The rate of working or power, P, is obtained from the product of the constant torque and the angular speed ω rad s^{-1}, thus

$$P = T\omega$$

but $\omega = 2\pi n$

where n is the angular speed in revolutions per second (rev s^{-1}), thus

$$P = 2\pi nT \tag{3.7}$$

If the torque T is in N m, the units for P are watts.

Example 3.3

The main drive-shaft of a racing-car engine is solid and has a diameter of 32 mm. What is the maximum power that may be transmitted at 100 rev s^{-1} if the shear stress in the shaft is not to exceed 80 MN m^{-2}?

From equation 3.6, the maximum torque is given by

$$T = \frac{\tau \text{ max}}{R} \frac{\pi R^4}{2}$$

thus $T = (80) \dfrac{\pi 16^3}{2}$ N mm = 515 N m

From equation 3.7 power, P, transmitted is given by

$$P = 2\pi \times 100 \times 515 \text{ W} = 323 \cdot 5 \text{ kW}$$

3.7 TORQUE AND ANGLE-OF-TWIST DIAGRAMS

So far we have considered the case of a shaft subjected to constant torque. Let us now look at the variation in torque in a uniform circular shaft, AB, rigidly restrained at each end and carrying a concentrated torque (or couple) T at some point C in the length (figure 3.7).

Reactant torques T_A and T_B will be generated at the ends of the shaft and to ensure equilibrium of torques we must have that

$$T = T_A + T_B \tag{a}$$

Since there are two unknown reactant torques and only one statical equation, the problem is statically indeterminate and we must consider deformations in order to reach a solution.

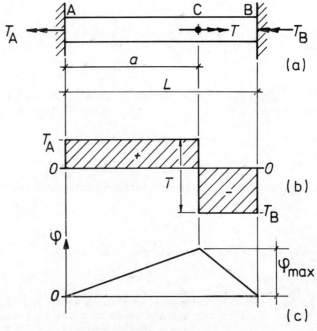

Figure 3.7

We adopt the convention that a torque vector pointing to the right denotes a positive torque and that positive torques give rise to positive twists. The angle of twist of end B relative to end A is thus

$$\phi_B = \frac{Ta}{GJ} - \frac{T_B L}{GJ}$$

But ϕ_B is zero since both ends of the shaft are rigidly restrained, thus

$$T_B = T \frac{a}{L} \tag{b}$$

and from equation a

$$T_A = T\left(1 - \frac{a}{L}\right) \tag{c}$$

The distribution of torque can be shown by a torque diagram, the ordinates of which represent the algebraic sum of the torques to the right of a point in the shaft. Such a torque diagram, for the shaft considered above, is shown in figure 3.7b.

The angle of twist at any point in the shaft may be found from the ordinate of an angle-of-twist diagram such as that shown in figure 3.7c. In figure 3.7b there is a discontinuity at the point of application of the torque; this is due to the assumption that the torque is concentrated. In practice the torque will be distributed over a finite length of shaft. If we assume this distribution to be uniform over a length $2b$, the resulting torque and angle-of-twist diagrams are as shown in figures 3.8a and b.

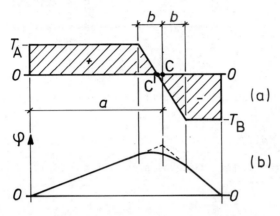

Figure 3.8

Suppose that the torque is distributed at an intensity m (per unit length of shaft), then

$$m \times 2b = T$$

or $m = \dfrac{T}{2b}$

The maximum angle of twist is less than that found for the case of the concentrated torque and it occurs at the point in the shaft where the torque is zero. This point is slightly to the left of C if $a > L/2$.

3.8   THE T/GJ DIAGRAM

Equation 3.5 applies to a circular-section shaft for which both the torque T and the polar second moment of area J may be functions of z. From this equation the rate of change of angle of twist along the shaft is

$$\frac{d\phi}{dz} = \frac{T}{GJ} \tag{3.8}$$

Thus if the ordinates of a torque diagram are divided by the torsional stiffness, GJ, of the shaft, we obtain the T/GJ diagram whose ordinates represent the rate of change of angle of twist.

Consider the T/GJ diagram for a circular shaft in which the torque and the polar second moment of area are functions of z, the distance along the shaft. Such a diagram is shown in figure 3.9.

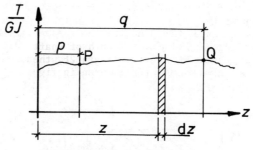

Figure 3.9

The area of a small element of the diagram is $(T/GJ)dz$, but from equation 3.8

$$d\phi = \frac{T}{GJ}\, dz$$

Integrating this expression between the limits P and Q in the shaft we obtain

$$\phi_Q - \phi_P = \int_P^q \frac{T}{GJ}\, dz$$

or $\phi_Q - \phi_P$ = area of the T/GJ diagram between P and Q

Hence the following theorem

The change in angle of twist between two points in a straight circular-section shaft is equal to the area of the T/GJ diagram between the same two points.

Applying the theorem between the ends A and B of the shaft in figure 3.7 we obtain.

$$\phi_B - \phi_A = \frac{T_A}{GJ}\, a - \frac{T_B}{GJ}\, (L - a)$$

but $\phi_B = \phi_A = 0$ and GJ is constant, thus

$$T_A = T_B \frac{L - a}{a}$$

Now $T = T_A + T_B = T_B \left(\frac{L-a}{a}\right) + T_B$

thus $T_B = T \dfrac{a}{L}$ and $T_A = T \left(1 - \dfrac{a}{L}\right)$ as before

The maximum twist in the shaft is at C where the torque changes sign. Applying the theorem between A and C, we have

$$\phi_C - \phi_A = \frac{T_A a}{GJ} = \frac{Ta}{GJ} \left(1 - \frac{a}{L}\right)$$

but $\phi_C = \phi_{max}$ and $\phi_A = 0$, thus

82

$$\phi_{max} = \frac{Ta}{GJ} \left(1 - \frac{a}{L}\right) \tag{a}$$

We now examine the more practical case shown in figure 3.8. The point of maximum twist in the shaft has now moved to C' where the torque is zero, thus

$$CC' = b\left(\frac{2a}{L} - 1\right)$$

and $\quad \phi_{C'} - \phi_A = \frac{T_A}{GJ}(a-b) + \frac{T_A}{GJ} b\left(1 - \frac{a}{L}\right)$

but $\phi_{C'} = \phi_{max}$ and $\phi_A = 0$, thus

$$\phi_{max} = \frac{Ta}{GJ} \left(1 - \frac{a}{L}\right)\left(1 - \frac{b}{L}\right) \tag{b}$$

This result should be compared with that obtained for the shaft under concentrated torque (equation a). Notice that both equations are identical if b = 0.

Example 3.4

The stepped shaft ABCD shown in figure 3.10a is made up of equal lengths of solid circular section. The polar second moments of area for sections AB, BC and CD are respectively $3J$, $2J$ and J. The shaft carries a uniformly distributed torque of m per unit length on the centre section BC. The ends A and D are fully restrained against

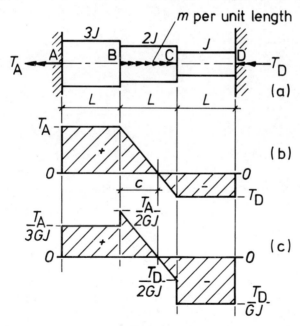

Figure 3.10

twisting. Determine the reactant torques at A and D and the greatest angle of twist in the shaft. The effect of stress concentrations at sudden changes of cross-section are to be neglected.

The torque diagram is shown in figure 3.10b. The T/GJ diagram in figure 3.10c may be constructed from the torque diagram by dividing ordinates by appropriate values of GJ.

Since the ends of the shaft do not twist, $\phi_A = \phi_D = 0$ and the total area of the T/GJ diagram must be zero, thus

$$\frac{T_A}{3GJ} L + \frac{T_A}{2GJ} \frac{c}{2} - \frac{T_D}{2GJ} \frac{(L - c)}{2} - \frac{T_D L}{GJ} = 0 \qquad (1)$$

where $\dfrac{c}{L} = \dfrac{T_A}{T_A + T_D}$

After substituting for c/L and rearranging, equation 1 yields the following quadratic for T_A/T_D

$$7\left(\frac{T_A}{T_D}\right)^2 - 8\left(\frac{T_A}{T_D}\right) - 15 = 0$$

thus $T_A = \dfrac{15}{7} T_D$ \qquad (2)

The positive root is taken since the sign of T_D has already been allowed for.

From consideration of equilibrium of torques we have the equation

$$T_A + T_D = mL \qquad (3)$$

Therefore, from equations 2 and 3

$$T_D = \frac{7}{22} mL$$

and $T_A = \dfrac{15}{22} mL$

The greatest angle of twist occurs where T/GJ is zero, thus

$$\phi_{max} - \phi_A = \frac{T_A}{3GJ} L + \frac{T_A}{2GJ} \frac{c}{2}$$

After substituting for c and T_A and noting that ϕ_A is zero, we obtain

$$\phi_{max} = \frac{665}{1936} \frac{mL^2}{GJ}$$

Example 3.5

A solid circular-section shaft AB of length L tapers uniformly from diameter $2D$ at A to D at B. The end A is rigidly restrained against twisting. The end B is unrestrained and carries a positive concentrated torque T. Determine the maximum angle of twist in the shaft and the maximum value of the shear stress.

Figure 3.11a shows the geometry of the shaft. At a particular section ZZ, distant z from A, the diameter d_z is given by

$$d_z = D\left(2 - \frac{z}{L}\right)$$

The polar second moment of area at ZZ is therefore

$$J_z = \frac{\pi D^4}{32}\left(2 - \frac{z}{L}\right)^4$$

The rate of change of angle of twist in the shaft, is thus

$$\frac{d\phi}{dz} = \frac{T}{GJ_z} = \frac{32T}{\pi GD^4}\frac{1}{(2 - z/L)^4} \tag{1}$$

The change in angle of twist between A and B is equal to the total area of the T/GJ diagram in figure 3.11b which has been constructed

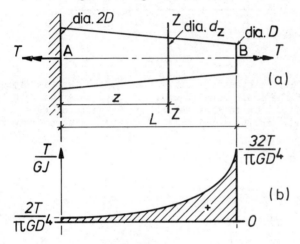

Figure 3.11

from equation 1. This area is obtained by integrating equation 1 between the limits $z = 0$ and $z = L$, thus

$$\phi_B - \phi_A = \frac{32T}{\pi GD^4}\int_0^L \frac{dz}{(2 - z/L)^4}$$

To perform the integration, we make the substitution

$$x = (2 - z/L)$$

Then, noting that ϕ_A is zero, we have

$$\phi_B = \frac{32TL}{\pi GD^4}\int_2^1 (-)\frac{dx}{x^4} = \frac{28TL}{3\pi GD^4}$$

The maximum shear stress occurs at end B and is given by

$$\tau_{max} = \frac{16T}{\pi D^3}$$

Example 3.6

A uniform solid circular-section shaft AB of length L is shown in figure 3.12a. The end A is rigidly restrained against rotation and the end B is attached to a torsion spring of stiffness K. The shaft carries a uniformly distributed torque of m per unit length. Determine the rotation of end B and the maximum rotation in the shaft.

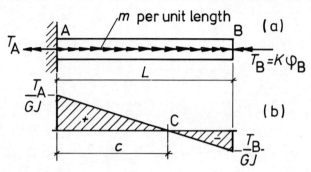

Figure 3.12

Because of the elastic restraint at B, the reactant torques at the ends of the shaft are unequal. To determine the magnitudes of these torques, we note that the change in angle of twist between A and B is equal to the total area of the T/GJ diagram (figure 3.12b).

$$\phi_B - \phi_A = \frac{T_A}{GJ}\frac{c}{2} - \frac{T_B}{GJ}\frac{(L-c)}{2}$$

Now $\dfrac{c}{L} = \dfrac{T_A}{T_A + T_B}$, $\phi_A = 0$ and $\phi_B = \dfrac{T_B}{K}$

therefore

$$T_A = T_B\left(1 + \frac{2GJ}{LK}\right) \tag{1}$$

For equilibrium of torques we have

$$T_A + T_B = mL \tag{2}$$

Thus from equations 1 and 2

$$T_B = \frac{mL^2K}{2(LK + GJ)}$$

$$T_A = \frac{mL(LK + 2GJ)}{2(LK + GJ)}$$

and $\quad \phi_B = \dfrac{mL^2}{2(LK + GJ)}$

The maximum angle of twist occurs at C where T/GJ is zero, thus

$$\phi_C - \phi_A = \frac{T_A}{GJ}\frac{c}{2}$$

86

since $\phi_C = \phi_{max}$ and $\phi_A = 0$

$$\phi_{max} = \frac{mL^2}{8GJ} \left(\frac{LK + 2GJ}{LK + GJ} \right)^2$$

3.9 PROBLEMS FOR SOLUTION

1. A steel shaft ABCD has a total length of 1·3 m made up as follows: AB = 300 mm, BC = 380 mm and CD = 620 mm. AB is hollow with outside diameter 100 mm and inside diameter d mm. BC and CD are solid having diameters of 100 mm and 89 mm respectively. If equal and opposite torques are applied to the ends of the shaft, find to the nearest mm, the maximum permissible value for d if the greatest shear stress in AB is not to exceed that in CD.

If the torque applied to the shaft is 9 kN m, what is the total angle of twist? $G = 81$ GN m^{-2}.
(73 mm, 1·16°)

2. A hollow shaft of circular cross-section is to have an internal diameter of half the outside diameter. It is to be designed to transmit 30 kW at 8 rev s^{-1} and the maximum shear stress is not to exceed 83 MN m^{-2}. $G = 81$ GN m^{-2}.

Calculate

(a) the external diameter of the hollow shaft
(b) the angle of twist over a length of 2 m
(c) the percentage difference in weight of the hollow shaft compared with a solid shaft designed for the same conditions.

(34 mm, 6·8°, 78%)

3. A composite shaft consists of a steel rod 25 mm diameter, encased in a gun-metal tube 25·01 mm internal diameter and 38 mm external diameter. The rod and the tube are rigidly attached to each other at their ends. The shaft is subjected to a torque of 624 N m. Calculate the maximum shear stress in the two materials and the angle of twist per unit length. G for steel = 80 GN m^{-2}, G for gun metal = 37 GN m^{-2}.
(For steel, $\tau_{max} = 67·8$ MN m^{-2}, for gun metal, $\tau_{max} = 47·6$ MN m^{-2}, angle of twist = 3·9° m^{-1})

4. A torsion test on a solid circular steel specimen of 22 mm diameter showed that the limit of elasticity in shear was reached when the torque was 280 N m.

Determine the diameter required, for a solid shaft made of the material used in the test, to transmit 37·3 kW at 6 rev s^{-1} with a maximum shear stress one-third of that found in the test. Find also the angle of twist per metre length of shaft. $G = 80$ GN m^{-2}.
(48·4 mm, 1·32° m^{-1})

5. A solid alloy shaft of 50 mm diameter is to be coupled in series with a hollow steel shaft of the same external diameter. Find the

internal diameter of the steel shaft if the angle of twist per unit length is to be 75% of that of the alloy shaft.

Determine the speed at which the shafts are to be driven to transmit 200 kW if the maximum shear stresses are to be limited to 54 and 77 MN m^{-2} in the alloy and steel respectively. G for steel = 2·2 × G for alloy.
(39·6 mm, 27·8 rev s^{-1})

6. A circular cross-section shaft AB of length L tapers non-uniformly from a radius R_0 at A. The loading on the shaft consists of a uniformly distributed torque of intensity m per unit length. End A of the shaft is rigidly restrained against twisting and end B is free from restraint. Derive a general expression for the shaft radius at a distance z from A if the angle of twist per unit length of shaft is to be constant.
$$\left(R_z = R_0 \left(1 - \frac{z}{L}\right)^{1/4}\right)$$

7. A uniform solid circular-section shaft of polar second moment of area J is rigidly restrained from twisting at its ends A and C. At a point B, distant L from A and $2L$ from C the shaft is cut at right angles to the longitudinal axis and a flexible coupling of stiffness K inserted. If the loading on the shaft consists of a uniformly distributed torque of intensity m per unit length running from A to B, determine the reactant torques and the relative angle of twist between the cut faces at B.
$$\left(T_A = \frac{mL}{2}\left(\frac{5LK + 2GJ}{3LK + GJ}\right), \quad T_C = \frac{mL^2K}{2(3LK + GJ)}, \quad \phi_{rel} = \frac{mL^2}{2(3LK + GJ)}\right)$$

8. A solid steel shaft ABCD is rigidly restrained against twisting at A and D. Sections AB, BC and CD are uniformly circular in cross-section and have radii 50 mm, 75 mm and 100 mm respectively. AB = 2 m, BC = 4 m and CD = 3 m. The shaft carries concentrated torques of + 60 kN m at B and - 150 kN m at C. Determine the reactant torques at A and D and the angles of twist at B and C.
G = 80 GN m^{-2}.
(T_A = - 10·5 kN m, T_D = + 100·5 kN m, ϕ_B = + 0·027 rad, ϕ_C = 0·023 rad)

4 BENDING

In chapter 1 we introduced the concept of the bending moment. The
effect of a moment is to produce curvature in a beam and to induce
a system of bending stresses in the cross-section. If the moment is
hogging, the top fibres of the beam will be in tension and the bottom
fibres in compression. There must therefore be a layer in the beam
where the stress due to the bending moment is zero. This layer is
called the neutral layer and a line indicating this layer in the
cross-section is called the neutral axis. In the theory that follows
we shall consider a bending moment acting alone. This situation, in
which both shear and axial forces are zero, is referred to as pure
bending.

4.1 THE SIMPLE THEORY OF BENDING

Consider the longitudinal element of a beam of uniform cross-section
shown in figure 4.1. The beam is in pure bending under a moment M.
It is assumed that the moment acts about an axis of symmetry of the
section so that the deformation is in the plane of the moment.
Suppose that the radius of curvature of the neutral layer before
bending is R_0 and after bending it deforms to a radius R. We shall
assume that the beam material is linearly elastic, that Young's
modulus is the same for tension and compression, that plane transverse sections remain plane and transverse after bending and that
all layers of the beam are free to expand or contract laterally.

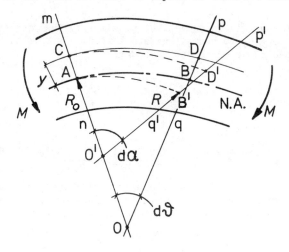

Figure 4.1

Due to bending, the plane pq rotates to p'q'. Let AB in figure
4.1 represent the neutral layer before bending and AB' the neutral

89

layer after bending. Since the neutral layer is not subject to stress, we have

$$AB = AB'$$

or $R_0 d\theta = R\, d\alpha$ (a)

since $AO = BO = R_0$ and $AO' = B'O' = R$.

Consider a layer of beam distance y above the neutral layer. Due to bending this layer deforms from CD to CD', thus the strain ε in the layer is given by

$$\varepsilon = \frac{(R + y)\, d\alpha - (R_0 + y)\, d\theta}{(R_0 + y)\, d\theta}$$

after substituting for $d\alpha$ from equation a

$$\varepsilon = \frac{R_0 y}{(R_0 + y)} \left[\frac{1}{R} - \frac{1}{R_0} \right]$$

The stress in layer CD is obtained by multiplying the strain by Young's modulus, thus

$$\sigma = \frac{E R_0 y}{(R_0 + y)} \left[\frac{1}{R} - \frac{1}{R_0} \right] \qquad\qquad\qquad\qquad (b)$$

Since no axial forces are present for pure bending, the sum of the stress-times-area products taken over the whole cross-section must be zero, thus

$$\int_A \sigma\, dA = E R_0 \left[\frac{1}{R} - \frac{1}{R_0} \right] \int_A \frac{y\, dA}{R_0 + y} = 0 \qquad\qquad (c)$$

At this stage it is convenient to simplify the problem by confining our attention to beams that are initially straight. Thus R_0 is infinite and equations b and c reduce to

$$\sigma = \frac{Ey}{R} \qquad\qquad\qquad\qquad\qquad\qquad\qquad (d)$$

and $\int_A y\, dA = 0$ (e)

Equation e implies that the neutral axis passes through the centroid (or centre of area) of the cross-section.

To satisfy the condition of moment equilibrium, the moments of all the stress-times-area products in the cross-section taken about the neutral axis must add up to the total moment M, thus

$$\int_A y\sigma\, dA = \frac{E}{R} \int_A y^2\, dA = M$$

or $\dfrac{EI}{R} = M$ (f)

where $I \left(= \int_A y^2\, dA \right)$ is the second moment of area of the cross-section.

From equations d and f we have the equations for bending of an
initially straight beam

$$\frac{M}{I} = \frac{\sigma}{y} = \frac{E}{R} \qquad\qquad\qquad (4.1)$$

4.1.1 The Second Moment of Area

We shall illustrate the calculation of second moment of area by
reference to rectangular and circular cross-sections. Rectangles
are of particular importance since they appear in many common
structural sections.

(a) *Rectangular section* Consider the rectangular cross-section of
breadth b and depth d shown in figure 4.2.

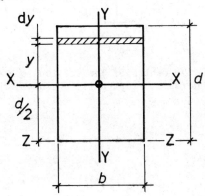

Figure 4.2

For bending in the vertical plane the neutral axis, XX, passes
through the centroid of the cross-section which by symmetry is at
half-depth, thus

$$I_{XX} = \int y^2 \, dA = \int_{-d/2}^{+d/2} y^2 b \, dy$$

therefore

$$I_{XX} = \frac{b}{3}\left(y^3\right)\Big|_{+d/2}^{+d/2} = \frac{bd^3}{12}$$

for bending in the horizontal plane about the neutral axis YY we
have, by similar reasoning

$$I_{YY} = \frac{db^3}{12}$$

To determine the second moment of area about an axis ZZ coinciding
with one of the shorter edges, we apply the parallel-axis theorem,
then

$$I_{ZZ} = (bd)\left(\frac{d}{2}\right)^2 + I_{XX} = \frac{bd^2}{3}$$

91

(b) *Circular section* Consider the circular cross-section of radius r shown in figure 4.3.

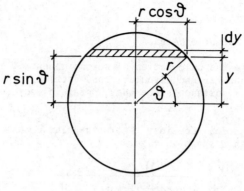

Figure 4.3

Any axis through the centre of the section is a neutral axis for bending in a plane normal to that axis. We have, therefore

$$I = \int y^2 \, dA$$

now $y = r \sin \theta$

hence

$$dy = r \cos \theta \, d\theta$$

also $dA = 2r \cos \theta \, dy = 2r^2 \cos^2 \theta \, d\theta$

thus $I = \displaystyle\int_{-\pi/2}^{+\pi/2} 2r^4 \sin^2 \theta \cos^2 \theta \, d\theta$

but $\sin^2 \theta \cos^2 \theta = \dfrac{1}{4} (1 - \cos^2 2\theta) = \dfrac{1}{8} (1 - \cos 4\theta)$

therefore

$$I = \frac{r^4}{4} \left(\theta - \frac{1}{4} \sin 4\theta \right)_{-\pi/2}^{+\pi/2} = \frac{\pi r^4}{4}$$

In terms of the diameter D $(= 2r)$

$$I = \frac{\pi D^4}{64}$$

4.1.2 The Elastic Section Modulus

From equation 4.1 above we have

92

$$\sigma = \frac{M}{I} \, y$$

thus the stress due to bending varies linearly across the section and becomes a maximum when y takes its maximum value. If the bending moment is hogging the maximum stress in tension occurs at the top surface of the beam and the maximum stress in compression occurs at the bottom surface. The two maximum values of stress will have the same numerical value if the neutral axis is also an axis of symmetry.

The quotient I/y is called the elastic section modulus and is given the symbol z, thus

$$\sigma \text{ (max)} = \frac{M}{z} \qquad\qquad\qquad\qquad\qquad\qquad\qquad \text{(h)}$$

Compare this equation with equation 2.1.

For a rectangular section

$$z_{XX} = \frac{bd^3}{12} \times \frac{2}{d} = \frac{bd^2}{6}$$

and $\quad z_{YY} = \frac{db^3}{12} \times \frac{2}{b} = \frac{db^2}{6}$

For a circular section

$$z = \frac{\pi d^4}{64} \times \frac{2}{d} = \frac{\pi d^3}{32}$$

Example 4.1

Determine the maximum and minimum elastic section moduli for the tee-section shown in figure 4.4. Sketch the bending-stress distribution for a sagging moment of 50 kN m. Bending is in the plane of the web.

The section has no axis of symmetry normal to the web so we must calculate the position of the centroid. Suppose that the centroid is at $\bar{y}$ from the foot of the web, then applying equation e we have

$$\int_{-\bar{y}}^{300-\bar{y}} y \times 15 \; dy + \int_{300-\bar{y}}^{325-\bar{y}} y \times 200 \; dy = 0$$

from which

$$\bar{y} = 235 \cdot 5 \text{ mm}$$

Alternatively, we may make use of the fact that the total moment of area of the section about the foot of the web must be equal to the sum of the individual moments of area of the two component rectangles about the same point, then

$$[(200)(25) + (300)(15)]\bar{y} = (300)(15)(150) + (200)(25)(312 \cdot 5)$$

hence $\bar{y} = 235 \cdot 5$ mm, as before.

We now require the second moment of area of the section about the neutral axis XX which passes through the centroid. The second moments of area of each of the component rectangles about their own

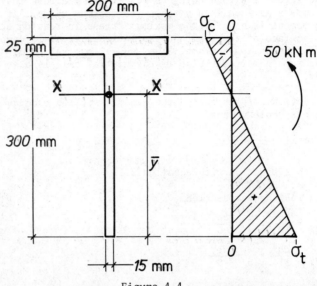

Figure 4.4

centroids are easily obtained. They are then referred to the neutral axis of the section by means of the parallel axis theorem, thus

$$I_{XX} = \frac{(15)(300)^3}{12} + (15)(300)(\bar{y} - 150)^2$$

$$+ \frac{(200)(25)^3}{12} + (200)(25)(312 \cdot 5 - \bar{y})^2 \text{ mm}^4$$

or $I_{XX} = 96 \cdot 55 \times 10^6 \text{ mm}^4$

There are two values for the elastic section modulus

$$Z \text{ (max)} = \frac{I_{XX}}{(325 - \bar{y})} = 1 \cdot 08 \times 10^6 \text{ mm}^3$$

$$Z \text{ (min)} = \frac{I_{XX}}{\bar{y}} = 0 \cdot 41 \times 10^6 \text{ mm}^3$$

The maximum stress in the section is obtained by dividing the applied moment by the *minimum* value of Z. Since the bending moment is sagging this stress will be tensile, then

$$\sigma_t = \frac{50 \times 10^6 \text{N mm}}{0 \cdot 41 \times 10^6 \text{ mm}^4} = 122 \text{ N mm}^{-2}$$

The largest compressive stress in the section is

$$\sigma_c = \frac{50 \times 10^6}{1 \cdot 08 \times 10^6} = 46 \cdot 3 \text{ N mm}^{-2}$$

94

Example 4.2

A wooden beam is 100 mm wide and 150 mm deep. A semi-circular groove of 20 mm radius is planed out of each side as shown in figure 4.5. The beam is simply supported over a span of 2 m and carries a uniformly distributed load of intensity $0 \cdot 4$ kN m^{-1} over the whole span, together with a point load of $0 \cdot 5$ kN at $0 \cdot 6$ m from one end. Determine the maximum stress in the beam.

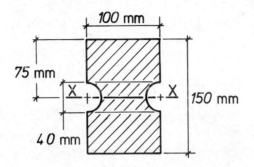

Figure 4.5

The neutral axis is clearly the horizontal axis of symmetry, XX. The second moment of area about XX is given by

$$I_{XX} = \frac{(100)(150)^3}{12} - \frac{\pi(40)^4}{64} = 28 \times 10^6 \text{ mm}^4$$

This calculation shows how a cut-out may be treated as having a negative second moment of area. In this case, the two semi-circles have the same second moment of area as a full circle of the same radius since their centres are both on the neutral axis.

Figure 4.6 shows the loading on the beam and the resulting shear-force and bending-moment diagrams. The end reactions are obtained in the usual way by the application of the equations of statical equilibrium.

The maximum bending moment occurs where the shear force is zero. If this point is x m from end B, we have from the shear-force diagram that

$$\frac{x}{0 \cdot 55} = \frac{1 \cdot 4 - x}{0 \cdot 01}$$

or $x = 1 \cdot 375$ m

therefore

$$M_{max} = - (1 \cdot 375)(0 \cdot 55) + \frac{(0 \cdot 4)(1 \cdot 375)^2}{2}$$

or $M_{max} = - 0 \cdot 378$ kN m

The elastic section modulus is given by

95

$$z = \frac{I}{y_{max}} = \frac{28 \times 10^6}{75} \text{ mm}^3$$

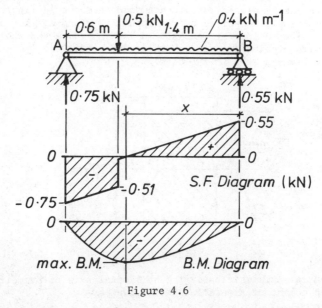

Figure 4.6

From equation h we have therefore that the maximum stress in the extreme fibres is given by

$$\sigma_b = \pm (0 \cdot 378 \times 10^6) \frac{75}{28 \times 10^6} = \pm 1 \cdot 0 \text{ N mm}^{-2}$$

Example 4.3

Determine the maximum stress due to bending in the cantilever of example 1.4. The cantilever section is shown in figure 4.7. Bending takes place about the major axis, XX.

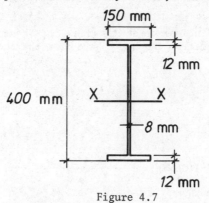

Figure 4.7

The cantilever section is symmetrical thus the centroid is in the middle of the web. Using the concept of negative second moment of area we have

$$I_{XX} = \frac{(150)(400)^3}{12} - \frac{(142)(376)^3}{12} \text{ mm}^4$$

thus

$$I_{XX} = 171 \times 10^6 \text{ mm}^4$$

and $Z_{XX} = 855 \times 10^3 \text{ mm}^3$

The maximum bending stress is at the wall where the bending moment takes its maximum value of 145 kN m, thus

$$\sigma_b(\text{max}) = \pm \frac{145 \times 10^6}{855 \times 10^3} = \pm 170 \text{ N mm}^{-2}$$

4.2 COMPOSITE BEAMS

It is sometimes necessary to construct beams of more than one material. Since, for example, it is difficult to obtain sound timber of large section, an equivalent-strength beam may be designed using a vertical steel web plate bolted between two narrow timber beams of the same depth. Alternatively two steel flange plates may be bolted to the top and bottom of a single timber beam. The steel and the timber then act compositely to carry the beam load.

Another example of composite-beam construction is the use of re-inforced concrete. Since concrete is not capable of carrying much tensile stress, steel reinforcing bars are embedded in the concrete on the tension side of the beam and are assumed to carry all the tensile stress. Additional reinforcement is sometimes used on the compression side to produce a doubly reinforced beam.

4.2.1 Timber Beams with Steel Web Plates

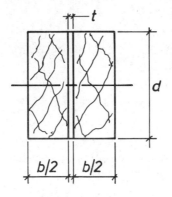

Figure 4.8

Figure 4.8 shows a steel web plate sandwiched between two timber beams. Under the action of a moment M, the composite beam will

97

bend into a circular arc of radius R. If the moments taken by the timber and steel are respectively M_t and M_s we have from equation 4.1

$$M = M_t + M_s = \frac{(EI)_t + (EI)_s}{R} \qquad (a)$$

It is possible to transform the section into an equivalent beam of one material. For the equivalent timber beam we have

$$M = \frac{E_t}{R}\left(I_t + \frac{E_s}{E_t} I_s\right) = \frac{E_t I_t'}{R} \qquad (b)$$

where I_t' is the second moment of area of an equivalent solid timber beam.

Referring to figure 4.8 and equation b we have

$$I_t' = \frac{bd^3}{12} + \left(\frac{E_s}{E_t}\right)\frac{td^3}{12}$$

or $\quad I_t' = \left(b + \frac{E_s}{E_t} t\right)\frac{d^3}{12} \qquad (c)$

The steel may be notionally transformed into timber by multiplying the thickness of the plate by the ratio of the Young's modulus for the steel to that for the timber.

In a similar way the composite beam may be transformed into one consisting entirely of steel. In this case the equivalent second moment of area is given by

$$I_s' = \left(\frac{E_t}{E_s} b + t\right)\frac{d^3}{12} \qquad (d)$$

The stress in the timber at a distance y from the neutral axis is given by

$$\sigma_t = \frac{M_t y}{I_t} \qquad (e)$$

but $\quad M_t = \frac{E_t I_t}{R}$ and $M = \frac{E_t I_t'}{R}$

thus eliminating E_t/R and substituting for M_t in equation e we have

$$\sigma_t = \frac{My}{I_t'} \qquad (f)$$

similarly for the stress in the steel we have

$$\sigma_s = \frac{My}{I_s'} \qquad (g)$$

Notice that the longitudinal strains at a particular distance from the neutral axis are the same in both steel and timber since the radius of curvature is the same for both materials.

Example 4.4

A composite beam consists of two timbers 150 mm wide and 300 mm deep on both sides of a steel plate 16 mm thick and 300 mm deep. Find the moment that may be carried by the beam if the tensile stress in the timber is limited to 7 MN m^{-2}. If the span of the beam is 3·6 m, what is the maximum uniformly distributed load that the beam will carry? E for steel = 200 GN m^{-2}, E for timber = 10 GN m^{-2}.

Transforming the composite beam into an equivalent timber beam we obtain a second moment of area given by

$$I_t' = 2\frac{(150)}{12}(300)^3 + \frac{(200)}{(10)}\frac{(16)}{12}(300)^3 \text{ mm}^4$$

or $I_t' = 1\cdot395 \times 10^{-3} \text{ m}^4$

If the timber stress is limited to 7 MN m^{-2} we have from equation f that

$$M = (7\cdot0)\frac{(1\cdot395 \times 10^{-3})}{(0\cdot150)} \text{ MN m}$$

thus $M = 65\cdot1$ kN m.

The maximum moment in a uniformly loaded beam is $WL/8$ where W is the total load and L is the span. If this moment is limited to that determined above, we have

$$W = \frac{8(65\cdot1)}{(3\cdot6)} = 144\cdot7 \text{ kN}$$

4.2.2 Timber Beams with Steel Flange Plates

Figure 4.9 shows a timber beam with steel flange plates secured with screws or bolts. The behaviour of such a beam in bending may be analysed by the transformation method dealt with in section 4.2.1. above.

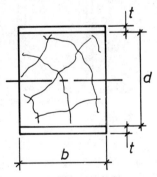

Figure 4.9

The behaviour of the two types of beam differs when shear forces are present. Whereas there is no relative shearing action between a steel web plate and the adjacent timber, in the case of the beam with flange plates, shear forces must be transmitted between the timber and the steel by means of the fasteners. To determine the shear force to be withstood by the fasteners it may be assumed that all the vertical shear on the beam is taken by the timber and furthermore that the resulting shear stress is uniformly distributed. If the vertical shear force is Q_V, the resulting uniform shear stress τ is given by

$$\tau = \frac{Q_V}{bd} \tag{a}$$

In order to preserve equilibrium a complementary shear stress (section 3.2) equal to τ must exist on the horizontal interface between the timber and the steel. The resulting horizontal shear force Q_h on a unit length beam is thus

$$Q_h = \tau \times b \times 1 = \frac{Q_V}{d} \tag{b}$$

Example 4.5

A composite steel and timber beam has a cross-section as shown in figure 4.9. The breadth, b, is 150 mm and the depth, d, of the timber is 200 mm. The thickness, t, of the steel plate is 3 mm. The beam is used to span 6 m and carries a concentrated load of 15 kN at 2 m from one support. Determine the maximum bending stresses in the steel and the timber.

If the screws used to attach the flanges to the timber are each capable of carrying a maximum shear force of 4·0 kN, how many will be necessary for the whole beam? E for steel = 200 GN m^{-2}, E for timber = 8 GN m^{-2}.

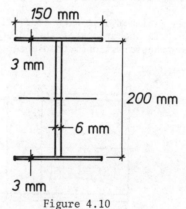

Figure 4.10

If the timber core is transformed into an equivalent steel web plate we obtain the section shown in figure 4.10 for which

$$I = \frac{(150)(206)^3}{12} - \frac{(144)(200)^3}{12} \text{ mm}^4$$

or $I = 13 \cdot 27 \times 10^{-6} \text{ m}^4$

100

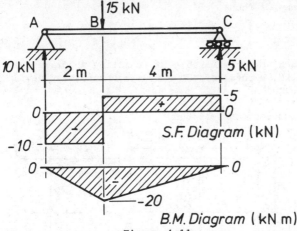

S.F. Diagram (kN)

B.M. Diagram (kN m)

Figure 4.11

From figure 4.11 the maximum bending moment in the beam is seen to be 20 kN m. Thus the maximum stress in the steel is given by

$$\sigma_S = \pm \frac{(20)(0 \cdot 106)}{(13 \cdot 27 \times 10^{-6})} \text{ kN m}^{-2}$$

or $\quad \sigma_S = \pm 160 \text{ MN m}^{-2}$

The longitudinal strain at a distance y from the neutral axis in the all steel beam is given by

$$\varepsilon = \frac{My}{E_S I}$$

At the junction of the web and flange the maximum value of this strain is

$$\varepsilon = \pm \frac{(20 \times 10^{-3})(0 \cdot 100)}{(200 \times 10^3)(13 \cdot 27 \times 10^{-6})} = \pm 7 \cdot 53 \times 10^{-4}$$

But this is also the maximum strain in the timber thus the maximum timber stress is

$$\sigma_t = \pm (7 \cdot 53 \times 10^{-4})(8 \times 10^3) = \pm 6 \cdot 0 \text{ MN m}^{-2}$$

Figure 4.11 also shows the shear-force diagram for the beam. From A to B the horizontal shear force intensity is

$$(\sigma_h)_{AB} = \frac{10}{0 \cdot 2} = 50 \text{ kN m}^{-1}$$

From B to C the horizontal shear force is

$$(\sigma_h)_{BC} = \frac{5}{0 \cdot 2} = 25 \text{ kN m}^{-1}$$

The total horizontal shear force to be resisted by the top and bottom screws is thus

$$(Q_h)_{total} = \left[(Q_h)_{AB}2 + (Q_h)_{BC}4\right] 2 = 400 \text{ kN}$$

Since each screw is capable of resisting a horizontal shear force of 4 kN, a total of 100 screws are necessary divided equally between lengths AB and BC.

4.2.3 Reinforced-concrete Beams

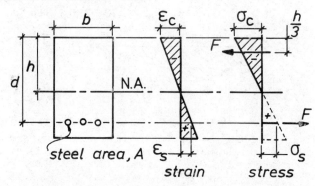

Figure 4.12

Figure 4.12 shows a concrete beam subjected to a bending moment that causes tension at the bottom surface and compression at the top. The beam is reinforced in the tension zone with three steel rods. Diagrams of the strain and stress distributions are also shown. The stress diagram is not exact since any tensile strength the concrete might have is neglected because it is very small in comparison with the compressive strength (see section 2.2).

Before we are able to determine the moment capacity of the beam, the position of the neutral axis must be found. If no axial forces act on the beam, the sum of the longitudinal forces due to bending must be zero, thus referring to figure 4.12 we have

$$F = \frac{hb\sigma_c}{2} = \sigma_s A \tag{a}$$

From the strain distribution we have

$$\frac{\varepsilon_s}{d - h} = \frac{\varepsilon_c}{h}$$

or $$h = \frac{d\varepsilon_c}{\varepsilon_c + \varepsilon_s} \tag{b}$$

but $\varepsilon_s = \dfrac{\sigma_s}{E_s}$ and $\varepsilon_c = \dfrac{\sigma_c}{E_c}$, thus

$$h = \frac{md\sigma_c}{\sigma_s + m\sigma_c} \tag{c}$$

where m (the modular ratio) = E_s/E_c.

Eliminating the stress ratio σ_s/σ_c between equations a and c we obtain the following relationship for the position of the neutral axis

$$h^2 + \frac{2Am}{b} h - \frac{2Amd}{b} = 0$$

from which the relevant solution for h is

$$h = \frac{Am}{b} \left[\left\{ 1 + \frac{2bd}{Am} \right\}^{\frac{1}{2}} - 1 \right] \qquad \text{(d)}$$

Referring to figure 4.12, the beam moment, M, is equal to the couple provided by the equal and opposite forces F. Since the compressive force F acts through the centroid of the triangular compressive stress diagram we have

$$M = F(d - \frac{h}{3}) \qquad \text{(e)}$$

The value of the moment is obtained by substitution from equation a for F and either equation c or d for h. The maximum moment capacity will depend on the limiting stresses that are permitted in the steel and the concrete.

Example 4.6

A rectangular reinforced-concrete beam is 200 mm wide and has an effective depth of 450 mm. It is reinforced only on the tension side with two steel bars each 600 mm^2 in area. Determine the maximum bending moment that the beam can support and the maximum stresses in the steel and concrete due to this loading. The allowable stresses in the concrete and the steel are 7·5 MN m^{-2} and 150 MN m^{-2} respectively. The modular ratio is 15.

From equation d we have

$$h = \frac{(2 \times 600)(15)}{(200)} \left\{ \left[1 + \frac{2(200)(450)}{(2 \times 600)(15)} \right]^{\frac{1}{2}} - 1 \right\} \text{ mm}$$

or $\quad h = 208 \cdot 5$ mm

From equation a the ratio of the steel stress to the concrete stress is

$$\frac{\sigma_s}{\sigma_c} = \frac{(208 \cdot 5)(200)}{2(2 \times 600)} = 17 \cdot 37$$

Since the ratio of the allowable stresses is 20, it is evident that the steel strength cannot be fully utilised. Setting σ_c as the full value of 7·5 MN m^{-2} we have

$$\sigma_c = - 7 \cdot 5 \text{ MN m}^{-2}$$

and $\quad \sigma_s = 130 \cdot 2$ MN m^{-2}

Thus again from equation a we have

103

$$F = (130 \cdot 2)(2 \times 600) \text{ N}$$

or $F = 156 \cdot 4 \text{ kN}$

The maximum moment capacity is then obtained from equation e thus

$$M = (156 \cdot 4)\left[450 - \frac{208 \cdot 5}{3}\right] \text{ N m}$$

or $M = 59 \cdot 5 \text{ kN m}$

Example 4.7

A concrete beam 250 mm wide by 450 mm deep is reinforced with two 16 mm diameter bars centred 50 mm from the top surface and with two 25 mm diameter bars centred 50 mm from the bottom surface. Determine the maximum sagging moment that the beam is capable of supporting if the permissible stresses of concrete and reinforcement are 7 and 140 MN m^{-2} respectively. Take $m = 15$.

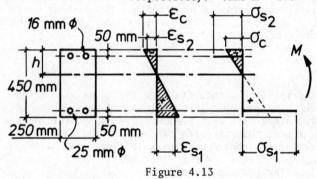

Figure 4.13

This problem concerns a doubly reinforced beam and must therefore be solved from first principles. The concepts that have been introduced in dealing with the singly reinforced beam still apply.

Referring to figure 4.13 in which subscripts s1 and s2 apply respectively to the tension and compression steel we have

$$\frac{\varepsilon_{s1}}{400 - h} = \frac{\varepsilon_{s2}}{h - 50} = \frac{\varepsilon_c}{h} \tag{1}$$

Since there are no axial forces acting

$$\sigma_{s1} \times 2\left(\frac{\pi 25^2}{4}\right) = \sigma_{s2} \times 2\left(\frac{\pi 16^2}{4}\right) + \frac{1}{2}\sigma_c(250)h$$

or $\sigma_{s1} = 0 \cdot 410\sigma_{s2} + 0 \cdot 127\sigma_c h$ \hfill (2)

From equation 1 we have

$$\frac{\sigma_{s1}}{400 - h} = \frac{\sigma_{s2}}{h - 50} = \frac{m\sigma_c}{h} \tag{3}$$

where $m = 15$.

104

Substituting for σ_{s1} and σ_{s2} in equation 2 we obtain

$$0 \cdot 127h^2 + 21 \cdot 15h - 6307 \cdot 5 = 0$$

from which

$$h = 154 \cdot 6 \text{ mm}$$

From equation 3

$$\frac{\sigma_{s1}}{\sigma_c} = 23 \cdot 8 \text{ and } \frac{\sigma_{s2}}{\sigma_c} = 10 \cdot 1$$

Since the ratio of the allowable steel to concrete stress is 20 it is evident that the attainment of the allowable stress in the tension steel is a limiting condition for the beam; thus if

$$\sigma_{s1} = 140 \text{ MN m}^{-2}$$

we have

$$\sigma_c = -5 \cdot 9 \text{ MN m}^{-2}$$

and $\sigma_{s2} = -59 \cdot 4$ MN m^{-2}

The moment capacity of the beam is thus given by

$$M = \sigma_{s1} \times 2 \left(\frac{\pi 25^2}{4}\right) (245 \cdot 4) + \sigma_{s2} \times 2 \left(\frac{\pi 16^2}{4}\right) (104 \cdot 6) + \frac{\sigma_c}{2} (250)(154 \cdot 6)^2 \, \frac{2}{3} \text{ N mm}$$

or $M = 48 \cdot 0$ kN m

4.3 COMBINED BENDING AND DIRECT STRESS

Certain structures are subject to combinations of axial load and bending. The stresses arising from such loads act in longitudinal planes and their combined effect may be determined quite simply by algebraic addition provided that the material remains linearly elastic.

Examples of structures that are loaded in this way are gravity dams, chimneys subject to wind and eccentrically loaded columns.

To illustrate the process of combining axial and bending stresses, we shall consider the short eccentrically loaded column of cross-sectional area A shown in figure 4.14.

The column carries a thrust P that is displaced from the centroid of the cross-section by a distance e. This eccentric load is statically equivalent to a thrust P applied at the centroid together with a moment Pe about the centroid. The stress distribution due to the substitution of this alternative loading system will not be affected provided we consider cross-sections that are not too near the top of the column. We are making use here of St Venant's principle which in formal terms states

If on some region of the boundary of a body in equilibrium,

105

the distribution of surface forces is altered, then the stress distributions at points in the body remote from this region are unaffected, provided that the alternative distribution is statically equivalent to the original distribution.

Thus if we consider the stress distribution at the base of the column, we have a constant stress of $-P/A$ due to the axial load and a bending stress varying linearly from Pe/Z to $-Pe/Z$, where Z is the elastic section modulus.

Some materials such as concrete, brickwork and cast iron are weak in tension and therefore it is of interest to determine the

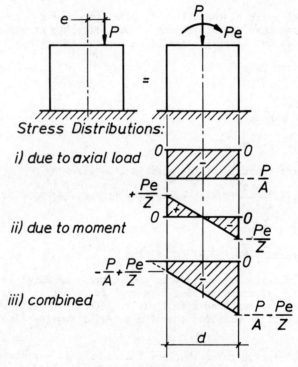

Figure 4.14

range of values of eccentricity that will not produce tension in the cross-section. If the column of figure 4.14 is rectangular with a depth d (in the plane of bending) and a breadth b, we have

$$Z = \frac{bd^3}{12} \times \frac{2}{d} = \frac{bd^2}{6}$$

thus for no tension in the column

$$-\frac{P}{A} + \frac{Pe}{Z} \leq 0$$

or $\quad e \leq \frac{Z}{A} \left(= \frac{d}{6} \right)$

106

Thus if the load is applied anywhere in the middle third of the depth no tension will be developed. Similarly for bending about the orthogonal axis no tension is developed if the load is placed within the middle third of the breadth. Consideration of a load positioned at some intermediate point between these axes enables the area of the cross-section to be determined within which the load may be placed to ensure no tension in the extreme fibres. This area is known as the core of the section. Figure 4.15 shows the section cores for a rectangular and a circular section.

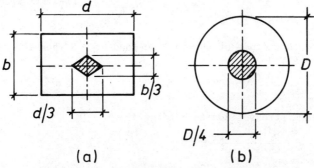

(a)   (b)

Figure 4.15

It should be noted that when both axial load and moment act on a section, the neutral axis (where the strain is zero) no longer passes through the centroid. Indeed it may be outside the section altogether as figure 4.14 shows.

Example 4.8

A short column has the cross-section shown in figure 4.16. A vertical load acts at P. If the maximum tensile stress in the column is limited to 30 MN m^{-2} find the magnitude of the applied load and the corresponding maximum compressive stress.

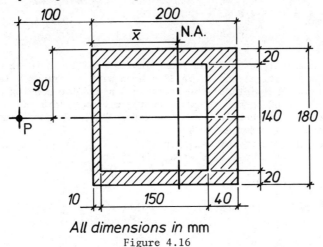

All dimensions in mm

Figure 4.16

107

To determine the position of the neutral axis we note that

$$(180 \times 200 - 150 \times 140)\bar{x} = (180)(200)(100) - (150)(140)(85)$$

thus $\bar{x} = 121$ mm.

The second moment of area about the neutral axis is given by

$$I_{NA} = \frac{(180)(200)^3}{12} + (180)(200)(21)^2 - \left[\frac{(140)(150)^3}{12} + (140)(150)(36)^2 \right] mm^4$$

or $\quad I_{NA} = 69 \cdot 3 \times 10^6$ mm^4

Let the magnitude of the applied load be P N then the direct stress is given by

$$\sigma_d = - \frac{P}{15000} \text{ N mm}^{-2}$$

the maximum stresses due to bending are

$$\sigma_{b1} = - \frac{P(100 + \bar{x})\bar{x}}{69 \cdot 3 \times 10^6} \text{ N mm}^{-2}$$

and $\quad \sigma_{b2} = + \frac{P(100 + \bar{x})(200 - \bar{x})}{69 \cdot 3 \times 10^6}$ N mm^{-2}

If the maximum tensile stress is limited to 30 N mm^{-2} we have

$$- \frac{P}{15000} + \frac{P(221)(79)}{69 \cdot 3 \times 10^6} = 30 \text{ N mm}^{-2}$$

hence $P = 162$ kN.

The maximum compressive stress is therefore given by

$$\sigma_{max} = - \frac{P}{15000} - \frac{P(221)(121)}{69 \cdot 3 \times 10^6} \text{ N mm}^{-2}$$

or $\quad \sigma_{max} = - 73 \cdot 4$ N mm^{-2}

Example 4.9

The trapezoidal masonry dam shown in figure 4.17 is to have its water face vertical and to be 33 m high with a maximum water height 3 m lower. The width of the crest is 4 m and the density of the masonry is 2500 kg m^{-3}. Determine the minimum width of the base if no tension is to develop when the water level is at maximum. Take the density of water as 1000 kg m^{-3}.

Since the density of water is 1000 kg m^{-3}, the maximum pressure (which is at the base of the dam) is given by

$$p_{max} = (1000)g(30) \text{ N m}^{-2}$$

where g is the acceleration due to the Earth's gravity (approximately $9 \cdot 81$ m s^{-2})

or $\quad p_{max} = 30g$ kN m^{-2}

For unit length of dam, the water-pressure force P which acts at a point 10 m above the base, is given by

$$P = \frac{30g}{2} (30)(1) = 450g \text{ kN}$$

The weight of the masonry per unit length is

$$W = (33) \frac{(4 + b)}{2} (1)(2500)g \text{ N}$$

or $W = 41 \cdot 25(4 + b)g \text{ kN}$

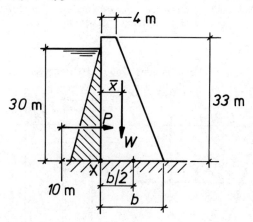

Figure 4.17

To determine the line of action of W we take moments of area for the dam cross-section about point X (see figure 4.17), then

$$W\bar{x} = \left[(33)(4)(2) + \frac{(33)}{2} (b - 4) \frac{(8 + b)}{3} \right] 2 \cdot 5g \text{ kN m}$$

hence

$$\bar{x} = \frac{(b^2 + 4b + 16)}{3(4 + b)}$$

The second moment of area of unit length of dam base about the centroid of the base is given by

$$I = \frac{1 \times b^3}{12} \text{ m}^4$$

If we assume that the moment about the centroid of the base produced by the water pressure exceeds that due to the masonry weight, the condition for no tension at X becomes

$$\frac{W}{b} \geq \left[10P - W\left(\frac{b}{2} - \bar{x}\right) \right] \frac{b}{2I}$$

since $b/2$ is always greater than $\bar{x}$. Hence

$$b^2 + 4b - 343 \cdot 3 > 0$$

109

or $b \geq 16 \cdot 64$ m

It is a simple matter to check that the above assumption con-
cerning the moments is upheld.

Example 4.10

A circular chimney is constructed from brick and mortar of density
2000 kg m^{-3}. The external and internal diameters are constant at
5 m and 4 m respectively. Determine the maximum height of the
chimney if no tension is to be induced at the base under a wind
pressure, assumed to be constant across the chimney width and acting
on the projected area of the chimney, given by

$$p = 0 \cdot 2 \sqrt{H} \text{ kN m}^{-2}$$

where H is the height above ground level in metres.

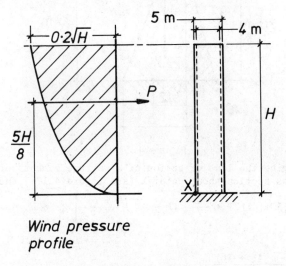

**Wind pressure
profile**

Figure 4.18

Figure 4.18 shows the wind-pressure profile which is parabolic
in form. From the geometry of a parabola (see appendix) we know
that the mean pressure is two-thirds of the maximum pressure.
Since the area of the chimney presented to the wind is constant,
the total wind force, P, acts at the centroid of the wind-pressure
profile (see appendix) or at a height of $5H/8$ above the ground.
The moment about the base of the chimney due to the wind is therefore given by

$$M = (0 \cdot 2 \sqrt{H}) \left(\frac{2}{3}\right) (5H) \left(\frac{5H}{8}\right) \text{ kN m}$$

or $M = \dfrac{5}{12} H^{5/2}$ kN m

The weight of the chimney is

110

$$W = \frac{\pi}{4} (5^2 - 4^2) H \times 2000 \times 9 \cdot 81 \text{ N}$$

or $W = 138 \cdot 6 \, H \text{ kN}$

For the chimney cross-section the area, A, and the second moment of area about the centroid, I, are given by

$$A = \frac{\pi}{4} (5^2 - 4^2) = \frac{9\pi}{4} \text{ m}^2$$

and $I = \frac{\pi}{64} (5^4 - 4^4) = \frac{369\pi}{64} \text{ m}^4$

The stress at point X (see figure 4.18) is therefore given by

$$\sigma_X = -\frac{W}{A} + \frac{M(2 \cdot 5)}{I}$$

if this stress is not to be tensile, we require that

$$\frac{W}{A} \geq \frac{M(2 \cdot 5)}{I}$$

or $\dfrac{(138 \cdot 6H)4}{9\pi} \geq \dfrac{5}{12} \dfrac{H^{5/2}(2 \cdot 5)64}{369\pi}$

hence

$$H^{3/2} \leq 341 \cdot 0$$

or $H \leq 48 \cdot 8 \text{ m}$

4.4 BENDING OF UNSYMMETRICAL BEAMS

The beam sections that we have been considering have had either one or two axes of symmetry and the maximum and minimum (or principal) second moments of area were relatively easy to determine. Some sections, such as the unequal angle, have no axis of symmetry and the determination of the principal second moments of area is more complex.

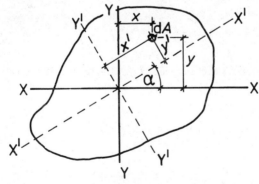

Figure 4.19

The simple theory of bending developed in section 4.1 applies if a moment acts about a principal axis. A moment applied about any other axis must be resolved into components acting about each of the principal axes. The final bending-stress distribution in the section is then the sum of the two distributions for bending about each principal axis.

Consider the arbitrarily shaped beam cross-section shown in figure 4.19. The perpendicular axes XX and YY pass through the centroid of the cross-section. Thus from the definition of the centroid we have

$$\int_A y \; dA = \int_A x \; dA = 0$$

where the integral is taken over the cross-section.

Suppose X'X' and Y'Y' are another pair of perpendicular axes which also pass through the centroid, then

$$\int_A y' \; dA = \int_A x' \; dA = 0$$

If X'X' makes an angle α with XX we have from figure 4.19 that

$$x' = x \cos \alpha + y \sin \alpha \qquad\qquad\qquad (a)$$

and $\quad y' = - x \sin \alpha + y \cos \alpha$

The second moment of area about axis X'X' is given by

$$I_{x'} = \int_A (y')^2 \; dA$$

or $\quad I_{x'} = \frac{1}{2} (I_x + I_y) + \frac{1}{2} (I_x - I_y) \cos 2\alpha - I_{xy} \sin 2\alpha \qquad (b)$

since

$$I_x = \int_A y^2 \; dA$$

$$I_y = \int_A x^2 \; dA$$

and we define a new section property, the product second moment of area, as

$$I_{xy} = \int_A xy \; dA$$

By similar reasoning

$$I_{y'} = \frac{1}{2} (I_x + I_y) - \frac{1}{2} (I_x - I_y) \cos 2\alpha + I_{xy} \sin 2\alpha \qquad (c)$$

From equations b and c we have

$$I_{x'} + I_{y'} = I_x + I_y$$

This result is to be expected since it is a corollary of the perpendicular-axis theorem.

The product second moment of area about axes X'X' and Y'Y' is given by

$$I_{x'y'} = \int_A x'y' \, dA$$

or $\quad I_{x'y'} = \frac{1}{2}(I_x - I_y) \sin 2\alpha + I_{xy} \cos 2\alpha \qquad$ (d)

If 2α is eliminated between equations b and d, we obtain the following equation for a circle

$$\left[I_{x'} - \frac{1}{2}(I_x + I_y)\right]^2 + (I_{x'y'})^2 = \frac{1}{2}(I_x - I_y)^2 + (I_{xy})^2$$

of radius $\left\{\left[\frac{1}{2}(I_x - I_y)\right]^2 + (I_{xy})^2\right\}^{\frac{1}{2}}$

and centre $\frac{1}{2}(I_x + I_y)$, 0

Figure 4.20 shows a plot of the circle of second moments of area.

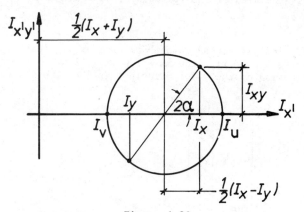

Figure 4.20

The principal second moments of area I_u and I_v occur when $I_{x'y'}$ is zero. From equation d we see that the major principal axis makes an angle α with the axis XX, where

$$\tan 2\alpha = \frac{2I_{xy}}{(I_y - I_x)} \qquad \text{(e)}$$

As an example of the method of calculation for the product second moment of area, consider the rectangular section shown in figure 4.21 whose centroid is displaced respectively distances p and q from axes OX and OY then $I_{xy} = \displaystyle\int_{q\,-\,b/2}^{q\,+\,b/2}\int_{p\,-\,d/2}^{p\,+\,d/2} xy\ dx\ dy$

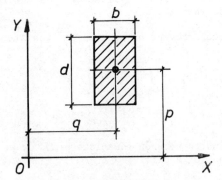

Figure 4.21

hence

$$I_{xy} = \frac{1}{2}\left[\left(q + \frac{b}{2}\right)^2 - \left(q - \frac{b}{2}\right)^2\right]\frac{1}{2}\left[\left(p + \frac{d}{2}\right)^2 - \left(p - \frac{d}{2}\right)^2\right]$$

or $\quad I_{xy} = qp(bd) = qpA$ \hfill (f)

where A is the area of the rectangular section.

Example 4.11

Find the principal second moments of area of an idealised unequal angle having the section 150 mm × 100 mm × 12 mm. Determine also the inclination of the major principal axis to the shorter leg.

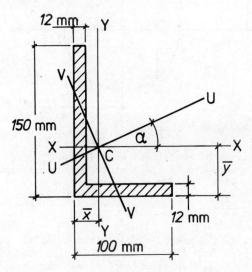

Figure 4.22

Figure 4.22 shows the cross-section of the angle with UU and VV representing the principal axes. The cross-sectional area A is given by

$$A = (100)(12) + (138)(12) = 2856 \text{ mm}^2$$

The position of the centroid is given by

$$\bar{x} = \frac{(12)(100)(50) + (12)(138)(6)}{(2856)} = 24 \cdot 5 \text{ mm}$$

and $\quad \bar{y} = \dfrac{(12)(150)(75) + (12)(88)(6)}{(2856)} = 49 \cdot 5 \text{ mm}$

To determine the second moments of area we make use of the fact that the second moment of area of a rectangle of depth d and breadth b about an edge perpendicular to the depth is given by $bd^3/3$ (see section 4.1.1), then

$$I_x = \frac{(12)(100 \cdot 5)^3}{3} + \frac{(100)(49 \cdot 5)^3}{3} - \frac{(88)(37 \cdot 5)^3}{3} \text{ mm}^4$$

or $\quad I_x = 6 \cdot 556 \times 10^6 \text{ mm}^4$

similarly

$$I_y = \frac{(12)(75 \cdot 5)^3}{3} + \frac{(150)(24 \cdot 5)^3}{3} - \frac{(138)(12 \cdot 5)^3}{3} \text{ mm}^4$$

or $\quad I_y = 2 \cdot 367 \times 10^6 \text{ mm}^4$

The product second moment of area may be determined from first principles in the following way

$$I_{xy} = \int_{-\bar{x}}^{100 - \bar{x}} \int_{-\bar{y}}^{12 - \bar{y}} xy \, dx \, dy + \int_{-\bar{x}}^{12 - \bar{x}} \int_{12-\bar{y}}^{150-\bar{y}} xy \, dx \, dy$$

thus

$$I_{xy} = \tfrac{1}{4}[(100 - \bar{x})^2 - (-\bar{x})^2][(12 - \bar{y})^2 - (-\bar{y})^2]$$

$$+ \tfrac{1}{4}[(12 - \bar{x})^2 - (-\bar{x})^2][(150 - \bar{y})^2 - (12 - \bar{y})^2] \text{ mm}^4$$

or $\quad I_{xy} = - 2 \cdot 296 \times 10^6 \text{ mm}^4$

Alternatively we may make use of the result given by equation f, thus

$$I_{xy} = (12)(138)[-(\bar{x} - 6)]\left(12 + \frac{138}{2} - \bar{y}\right)$$

$$+ (100)(12) \ [-(\bar{y} - 6)](50 - \bar{x}) \text{ mm}^4$$

or $\quad I_{xy} = - 2 \cdot 296 \times 10^6 \text{ mm}^4$ as before

115

From equation e we have

$$\tan 2\alpha = \frac{2\ (-\ 2 \cdot 296)\ 10^6}{(2 \cdot 367 - 6 \cdot 556)\ 10^6} = 1 \cdot 096$$

thus $2\alpha = 47 \cdot 63°$ or $\alpha = 23 \cdot 8°$.

From the circle diagram in figure 4.23 we obtain the principal second moments of area as

$$I_u = 7 \cdot 57 \times 10^6 \text{ mm}^4$$

$$I_v = 1 \cdot 35 \times 10^6 \text{ mm}^4$$

Notice that the *numerical* value of I_{xy} is used in the circle diagram which is then drawn in the manner shown in figure 4.20.

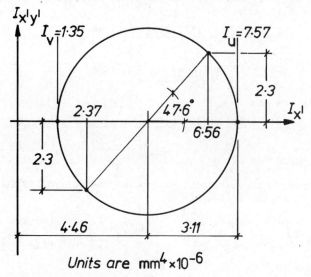

Units are $\text{mm}^4 \times 10^{-6}$

Figure 4.23

Example 4.12

A short length of the unequal angle of example 4.11 is subjected to a moment of 80 kN m in a plane perpendicular to the axis XX. Determine the maximum tensile and compressive stresses in the section and the position of the neutral axis. The moment vector is directed to the left if the section is positioned as shown in figure 4.22.

We begin by resolving the applied moment in the directions of the principal axes as shown in figure 4.24.

Consider a point Z distant u from the VV axis and v from the UU axis. The longitudinal stress at Z is given by

$$\sigma_z = -\frac{M_v u}{I_v} - \frac{M_u v}{I_u} \tag{1}$$

116

since both moment components produce compression at Z. Substituting numerical values, we have

$$\sigma_Z = - \frac{(3 \cdot 23) 10^6}{(1 \cdot 35 \times 10^6)} u - \frac{(7 \cdot 32) 10^6}{(7 \cdot 57 \times 10^6)} v \ \text{N mm}^{-2}$$

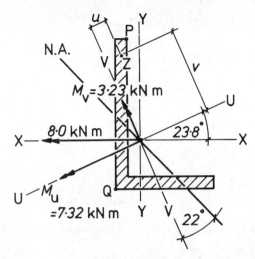

Figure 4.24

On the neutral axis, which must pass through the centroid, the stress σ_Z is zero, thus the equation of this axis is

$$- \frac{(3 \cdot 23)(7 \cdot 57)}{(1 \cdot 35)(7 \cdot 32)} u - v = 0$$

or $2 \cdot 474u + v = 0$

If the neutral axis is inclined at an angle β to the axis VV we have

$$\tan \beta = \frac{u}{v} = \frac{-1}{2 \cdot 474} = - 0 \cdot 404$$

thus $\beta = - 22 \cdot 0°$

The position of the neutral axis is indicated in figure 4.24. By inspection, the maximum compressive stress will occur at the corner P while the maximum tensile stress will be at the corner Q. On the XY coordinate system, the coordinates of P and Q are as follows

P, (- 12·5 mm, 100·5 mm)

and   Q, (- 24·5 mm, - 49·5 mm)

Transferring to the UV coordinate system we have, for point P (see equations a)

$$u_P = x \cos \alpha + y \sin \alpha$$

117

or $\quad u_P = (- 12\cdot5) \cos 23\cdot8° + (100\cdot5) \sin 23\cdot8°$

$\qquad\quad = + 29\cdot1 \text{ mm}$

and $\quad v_P = - x \sin \alpha + y \cos \alpha$

or $\quad v_P = -(- 12\cdot5) \sin 23\cdot8° + (100\cdot5) \cos 23\cdot8°$

$\qquad\quad = + 97\cdot0 \text{ mm}$

Similarly for point Q

$\quad u_Q = (- 24\cdot5) \cos 23\cdot8° + (- 49\cdot5) \sin 23\cdot8°$

$\qquad\quad = - 42\cdot4 \text{ mm}$

and $\quad v_Q = -(- 24\cdot5) \sin 23\cdot8° + (- 49\cdot5) \cos 23\cdot8°$

$\qquad\quad = - 35\cdot4 \text{ mm}$

Thus from equation 1 above

$$\sigma_P = - \frac{(3\cdot23)10^6(29\cdot1)}{(1\cdot35 \times 10^6)} - \frac{(7\cdot32)10^6(97\cdot0)}{(7\cdot57 \times 10^6)} = - 163\cdot4 \text{ N mm}^{-2}$$

and $\quad \sigma_Q = + \dfrac{(3\cdot23)10^6(42\cdot4)}{(1\cdot35 \times 10^6)} + \dfrac{(7\cdot32)10^6(35\cdot4)}{(7\cdot57 \times 10^6)}$

$\qquad\quad = + 135\cdot7 \text{ N mm}^{-2}$

4.5 PROBLEMS FOR SOLUTION

1. A cantilever beam of length 1·3 m is to carry a concentrated load of 10 kN at the free end. The cross-section of the beam is rect-angular with the breadth equal to half the depth. If the maximum stress due to bending is not to exceed 160 N mm^{-2}, determine the minimum depth of beam required.
(100 mm)

2. A steel beam having the section shown in figure 4.28 is used as a simply supported beam over a span of 1·22 m. The beam carries

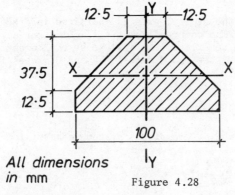

All dimensions in mm

Figure 4.28

118

three equal loads, one at midspan and one at each of the quarter span points. Determine the magnitude of the loads if the maximum stress due to bending is not to exceed 125 MN m^{-2}.
(4·15 kN)

3. A beam ABCD is simply supported at A and C. A concentrated load of 40 kN acts at B and a uniformly distributed load of intensity 60 kN m^{-1} runs from C to D. AB = 3 m, BC = 3 m and CD = 1 m. The beam has an I-cross-section, of depth 200 mm and breadth 100 mm with flanges and web 20 mm and 10 mm thick respectively. Determine the maximum numerical values of the shear force and bending moment and the maximum stress due to bending.
(60 kN, 45 kN m, 125 N mm^{-2})

4. A sandwich panel with a total thickness of 22 mm has aluminium skins and a central core of expanded plastic. One aluminium skin is 2 mm thick and the other is 5 mm thick. Determine the maximum moment per metre width that the panel may carry if the stress in the aluminium is limited to 100 MN m^{-2}. There is no limit to the stress in the plastic. Take E for aluminium and plastic as 75 and 1 GN m^{-2} respectively.
(3·6 kN m)

5. A rectangular beam 40 mm deep by 20 mm wide is made of a material for which Young's modulus in compression is 5% more than the value in tension. Determine the maximum moment capacity of the section, if the maximum tensile stress is not to exceed 80 MN m^{-2}. What is then the maximum compressive stress?
(432 N m, 82·0 MN m^{-2})

6. A composite beam consists of a steel I-section 400 mm deep, firmly connected to the underside of a concrete slab 600 mm wide and 150 mm deep. Calculate the permissible moment on the beam if the concrete stress is not to exceed 7 MN m^{-2}. The modular ratio is 15. The steel beam has a cross-sectional area of 10000 mm^2 and a major second moment of area of 312 × 10^6 mm^4.
(258 kN m)

7. A singly reinforced concrete beam has an effective depth of 420 mm. The steel area is 1500 mm^2. If the beam is designed for a moment of 81 kN m, determine the required breadth of beam on the assumption that both steel and concrete reach their limiting stresses of 150 and 7·5 MN m^{-2} respectively. The modular ratio is 15.
(334 mm)

8. A rectangular hollow steel section has outside dimensions 305 mm × 203 mm with a thickness of 6·35 mm. The section is used as a short column, the centre being completely filled with concrete. The column is subjected to an eccentric point load producing bending about the major axis only. If there are to be no tensile stresses in the concrete, determine the maximum eccentricity of the load. Take the modular ratio as 16.
(76 mm)

9. A trapezoidal masonry dam has a height of 20 m. The water face

is sloping and is inclined to the vertical at an angle θ, where tan θ = 0·2. The outside face is vertical and the water level is coincident with the top of the dam. Determine the minimum crest width if no tensile stresses are to be induced in the masonry. The ratio of the densities of masonry and water is 2·3.
(13·2 m)

10. A tie-bar of rectangular section 80 mm × 30 mm sustains an axial load of 100 kN. If the line of action of the load does not change, what depth of metal may safely be removed from one of the narrow sides in order that the maximum stress over the reduced width may not exceed 100 MN m^{-2}?
(18 mm)

11. An unequal angle section 229 mm by 102 mm has a thickness of 19 mm. Estimate the values of the principal second moments of area.
(33×10^6 mm^4, $2·7 \times 10^6$ mm^4)

12. A zed-section beam has the dimensions shown in figure 4.29. Determine the bending stress at point X due to a moment of 5·5 N m in a horizontal plane through the centroid.
(15·6 N mm^{-2})

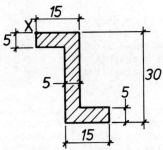

All dimensions in mm

Figure 4.29

5 DEFLEXION OF BEAMS

In chapter 4 we were able to investigate bending stresses in beams using the first part of equation 4.1. This equation may also be used to examine beam deflexions.

5.1 THE DEFLEXION EQUATION

From equation 4.1 we have the following relationship between bending moment, M, and radius of curvature, R

$$M = EI \frac{d^2 y}{dx^2} \Big/ \left[1 + \left(\frac{dy}{dx} \right)^2 \right]^{3/2} \tag{a}$$

When the slope (dy/dx) is small the reciprocal of the radius of curvature (the curvature) is given approximately by

$$\frac{1}{R} = \frac{d^2 y}{dx^2} \tag{b}$$

where x represents a distance along the beam and y is the corresponding vertical deflexion. Figure 5.1 shows a loaded beam with a positive (downward) deflexion. The origin of the coordinates x and y is at the left-hand support. From equations a and b we have for small deflexions

$$M = EI \frac{d^2 y}{dx^2} \tag{5.1}$$

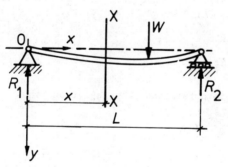

Figure 5.1

The sign convention assumed here is that both downward deflexions and hogging moments (producing tension at the top of the beam) are positive.

Using this convention, the moment to the left of section XX in figure 5.1 is given by

$$M_x = -R_1 x$$

The bending moment in a beam subjected to lateral loading only may therefore be expressed as a function of x. Thus from equation 5.1

Curvature $\dfrac{d^2y}{dx^2} = \dfrac{M_x}{EI}$

Slope $\dfrac{dy}{dx} = \displaystyle\int \dfrac{M_x}{EI}\, dx$

Deflexion $y = \displaystyle\iint \dfrac{M_x}{EI}\, dx\, dx$

Example 5.1

Determine the deflected form of a simply supported uniform beam of length L under a uniformly distributed load of w. What is the slope at the ends of the beam and the deflexion at the centre?

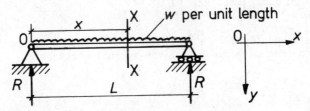

Figure 5.2

Referring to figure 5.2 we have the bending moment at XX as

$$M_x = -Rx + wx\left(\frac{x}{2}\right)$$

The total distributed load between O and XX is wx and its line of action lies half-way between these points.

From equation 5.1

$$EI\,\frac{d^2y}{dx^2} = -Rx + \frac{wx^2}{2} \qquad (1)$$

Integrate once to obtain the slope equation, thus

$$EI\,\frac{dy}{dx} = \frac{-Rx^2}{2} + \frac{wx^3}{6} + A \qquad (2)$$

The term EI does not need to be included in the integration since for this example it is constant; A is a constant of integration.

Integrate again to obtain the deflexion equation

$$EIy = -\frac{Rx^3}{6} + \frac{wx^4}{24} + Ax + B \qquad (3)$$

B is another constant of integration. Both A and B may be evaluated by taking account of the boundary conditions. For this beam we note that the deflexion must be zero at the ends of the beam, that is, when $x = 0$ and L; for $x = 0$, $y = 0$ and $B = 0$; for $x = L$, $y = 0$; thus

122

$$- \frac{RL^3}{6} + \frac{wL^4}{24} + AL = 0$$

or $\quad A = \frac{RL^2}{6} - \frac{wL^3}{24}$

We know from the condition of vertical equilibrium that

$$R = \frac{wL}{2}$$

thus $A = \dfrac{wL^3}{24}$

hence from equation 3

$$y = \frac{wL^4}{24EI} \left[\left(\frac{x}{L}\right)^3 - 2\left(\frac{x}{L}\right)^2 + 1 \right] \left(\frac{x}{L}\right)$$

The final slope equation is obtained by substitution for R and A, in equation 2 above, then, using the symbol θ for slope, we have

$$\frac{dy}{dx} = \theta = \frac{wL^3}{2EI} \left[4\left(\frac{x}{L}\right)^3 - 6\left(\frac{x}{L}\right)^2 + 1 \right]$$

For the slopes at the ends of the beam

$$x = 0, \; \theta = \theta_A = \frac{wL^3}{24EI}$$

$$x = L, \; \theta = \theta_B = \frac{wL^3}{24EI} [4 - 6 + 1] = \frac{- wL^3}{24EI}$$

where clockwise rotations are positive.

The deflexion at midspan is obtained by setting $x = L/2$ in the expression for y

$$y_{max} = \frac{wL^4}{24EI} \left(\frac{1}{2}\right) \left(\frac{1}{2^3} - \frac{2}{2^2} + 1\right) = \frac{5wL^4}{384EI}$$

Two further simple examples will now be given without detailed explanation.

Example 5.2

A uniform cantilever of length L has a concentrated load W at the free end. Determine the equations for slope and deflexion and values for these at the load point.

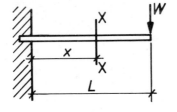

Figure 5.3

Referring to figure 5.3, the bending moment at XX is

$$M_x = + W(L - x)$$

From equation 5.1

$$EI \frac{d^2y}{dx^2} = W(L - x)$$

therefore

$$EI \frac{dy}{dx} = WLx - \frac{Wx^2}{2} + A$$

and $\quad EIy = \frac{WLx^2}{2} - \frac{Wx^3}{6} + Ax + B$

Now at the wall, $x = 0$, $dy/dx = 0$ and $y = 0$, thus $A = B = 0$, hence

$$\frac{dy}{dx} = \theta = \frac{WL^2}{2EI} \left(\frac{x}{L}\right) \left[2 - \left(\frac{x}{L}\right) \right]$$

and $\quad y = \frac{WL^3}{6EI} \left(\frac{x}{L}\right)^2 \left[3 - \left(\frac{x}{L}\right) \right]$

At the free end, $x = L$ thus

$$\theta_{max} = \frac{WL^2}{2EI}$$

and $\quad y_{max} = \frac{WL^3}{3EI}$

Example 5.3

A uniform cantilever of length L has a load W uniformly distributed along its length. Determine the equations for slope and deflexion and give values for these at the free end.

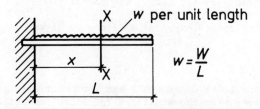

Figure 5.4

From figure 5.4

$$M_x = \frac{w}{2} (L - x)^2$$

thus

$$EI \frac{d^2y}{dx^2} = \frac{w}{2} (L^2 - 2Lx + x^2)$$

124

$$EI \frac{dy}{dx} = \frac{w}{2} \left[L^2 x - Lx^2 + \frac{x^3}{3} \right] + A$$

and $EIy = \dfrac{w}{2} \left(\dfrac{L^2 x^2}{2} - \dfrac{Lx^3}{3} + \dfrac{x^4}{12} \right) + Ax + B$

At $x = 0$, $dy/dx = 0$ and $y = 0$, thus $A = B = 0$, therefore

$$\frac{dy}{dx} = \theta = \frac{wL^3}{6EI} \left(\frac{x}{L} \right) \left[3 - 3 \left(\frac{x}{L} \right) + \left(\frac{x}{L} \right)^2 \right]$$

and $y = \dfrac{wL^4}{24EI} \left(\dfrac{x}{L} \right)^2 \left[6 - 4 \left(\dfrac{x}{L} \right) + \left(\dfrac{x}{L} \right)^2 \right]$

At the free end, $x = L$ thus

$$\theta_{max} = \frac{wL^3}{6EI} = \frac{WL^2}{6EI}$$

and $y_{max} = \dfrac{wL^4}{8EI} = \dfrac{WL^3}{8EI}$

5.2 SUPERPOSITION

An elastic beam carrying a combination of loads may be analysed using the principle of superposition. This permits the addition of separate slopes and deflexions due to a number of simple loading systems. The principle applies only if the structure remains linearly elastic, since then solutions of differential equations of the form given in equation 5.1 may be added.

As an example of superposition, the slope and deflexion at the free end of a cantilever carrying the combined loadings shown in examples 5.2 and 5.3 above would be

$$\theta_{max} = \frac{WL^2}{2EI} + \frac{WL^2}{6EI} = \frac{2WL^2}{3EI}$$

and $y_{max} = \dfrac{WL^2}{3EI} + \dfrac{WL^3}{8EI} = \dfrac{11WL^3}{24EI}$

Example 5.4

A horizontal cantilever of length $3a$ carries two concentrated loads: W at a from the fixed end and P at a from the free end. Obtain an expression for the deflexion at the free end due to this loading.

Refer to figure 5.5, then due to the load W alone

$$y_B = \frac{Wa^3}{3EI}$$

and $\theta_B = \dfrac{Wa^2}{2EI}$

thus

$$y_{D_1} = y_B + 2a\theta_B$$

or $\quad y_{D_1} = \dfrac{Wa^3}{3EI} + \dfrac{Wa^3}{EI} = \dfrac{4}{3}\dfrac{Wa^3}{EI}$

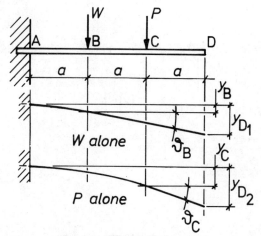

Figure 5.5

Due to the load P alone

$$y_C = \frac{P(2a)^3}{3EI} = \frac{8Pa^3}{3EI}$$

and $\quad \theta_C = \dfrac{P(2a)^2}{2EI} = \dfrac{2Pa^2}{EI}$

Thus $\quad y_{D_2} = y_C + a\theta_C$

or $\quad y_{D_2} = \dfrac{8Pa^3}{3EI} + \dfrac{2Pa^3}{EI} = \dfrac{14Pa^3}{3EI}$

Then $\quad y_D = y_{D_1} + y_{D_2}$

or $\quad y_D = \dfrac{2a^3}{3EI}\left[2W + 7P\right]$

5.3 PURE BENDING

The calculation of deflexions under pure bending conditions is particularly easy. Pure bending implies that the moment is constant along the beam, thus for a beam of uniform section, the radius of curvature is constant (see equation a of section 5.1) and the beam is bent into the arc of a circle.

Example 5.5

A beam of I-section is 200 mm deep and has flanges 100 mm wide by 10 mm thick. The web is 7·5 mm thick. It is simply supported over a length of 3·35 m and overhangs each support by 1·07 m. Two concentrated loads of equal value are carried at each end of the beam.

126

If the deflexion at mid-span is 6·35 mm above the supports, find the loads and the maximum stress due to bending. Neglect the weight of the beam and take $E = 200$ GN m^{-2}.

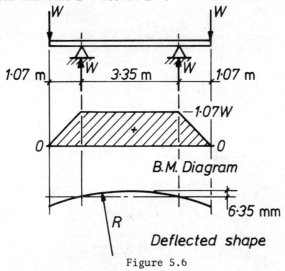

Figure 5.6

The relevant second moment of area is given by

$$I = \frac{100 \times 200^3}{12} - \frac{92 \cdot 5 \times 180^3}{12} \text{ mm}^4$$

thus

$$I = 21 \cdot 67 \times 10^{-6} \text{ m}^4$$

and $EI = 4 \cdot 34$ MN m

Referring to figure 5.6, the radius of curvature between the supports is

$$R = \frac{EI}{M} = \frac{4 \cdot 05}{W} \text{ m (if } W \text{ is in MN)} \tag{1}$$

From the properties of the chords of a circle we have

$$\left(\frac{3 \cdot 35}{2}\right)^2 = \left(2R\right)\left(\frac{6 \cdot 35}{1000}\right)$$

since the mid-span deflexion is small compared with R. Substituting for R from equation 1 we obtain

$$W = 18 \cdot 33 \times 10^{-3} \text{ MN}$$

The maximum bending stress, σ, between the supports is constant and is given by

$$\sigma = \frac{My}{I} = \frac{(1 \cdot 07 \times 18 \cdot 33 \times 10^{-3}) \, 100 \times 10^{-3}}{21 \cdot 67 \times 10^{-6}} \text{ MN m}^{-2}$$

127

thus $\sigma = 90 \cdot 6$ MN m^{-2}

5.4 BENDING MOMENTS HAVING A DISCONTINUOUS FIRST DERIVATIVE - THE UNIT FUNCTION

The moment expressions in the examples we have met so far have been functions of x that were easy to integrate. Many cases of beam loading however, give rise to discontinuities in the first derivative of the bending moment (i.e. the shear-force) and need special attention when the moment integral is to be determined. Consider the beam shown in figure 5.7.

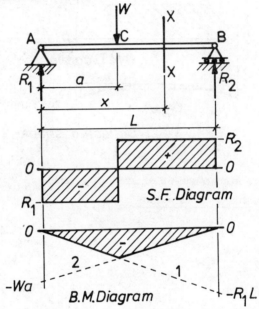

Figure 5.7

When the section XX lies between A and C (i.e. $0 < x < a$) we have

$$Q_x = \frac{dM_x}{dx} = -R_1$$

thus $M_x = -R_1 x + C_1$

since $M_x = 0$ when $x = 0$, we have $C_1 = 0$ and

$$M_x = -R_1 x \qquad \text{(a)}$$

When XX lies between C and B ($a < x < L$) we have

$$Q_x = \frac{dM_x}{dx} = -R_1 + W$$

thus $M_x = (-R_1 + W)x + C_2$

128

since $M_x = - R_1a$ when $x = a$, we have $C_2 = - Wa$ and

$$M_x = - R_1x + W[x - a] \qquad (b)$$

On the bending-moment diagram in figure 5.7 equation a is represented by line 1 and equation b by line 2.

The integration of the two moment expressions could be carried out in two stages (from A to C and then from C to B, ensuring continuity of slope and deflexion at C), but this is tedious. Instead it is possible to write down a single expression for the bending moment using a unity function.

The unity function, f_u, is simply a function that can take the value zero or unity depending on the value of x.

The complete bending-moment expression for the beam in figure 5.7 is thus

$$M_x = - R_1x + f_uW[x - a] \qquad (c)$$

where $f_u = 0$ for $0 < x < a$ and $f_u = 1$ for $a < x < L$. Thus

$$EI \frac{d^2y}{dx} = - R_1x + f_uW[x - a]$$

and $\quad EI \frac{dy}{dx} = \frac{- R_1x^2}{2} + \int f_uW[x - a] \; dx + A$

If the range of integration extends from $x = 0$ to $x \leq a$ then $f_u = 0$ and

$$EI \frac{dy}{dx} = - \frac{R_1x^2}{2} + A \qquad (d)$$

On the other hand, if the range of integration extends to an upper limit of $x > a$, we have

$$\int f_uW[x - a] \; dx = \int W[x - a] \; d \; (x - a)$$

and $\quad EI \frac{dy}{dx} = \frac{- R_1x^2}{2} + \frac{W}{2} [x - a]^2 + A \qquad (e)$

Similarly

$$EI_y = \frac{- R_1x^3}{6} + Ax + B \text{ for } x \leq a \qquad (f)$$

and $\quad EI_y = \frac{- R_1x^3}{6} + \frac{W}{6} [x - a]^3 + Ax + B \text{ for } x > a \qquad (g)$

The end result is as if the unity function in c above had been left out and the term in square brackets integrated *as a whole*. If we further stipulate that terms in square brackets are to be ignored (i.e. set to zero) when they become negative, we obtain equation d

129

from equation e and equation f from equation g. Thus equations e and g are the general equations for the problem.

5.5 MACAULAY'S METHOD

Macaulay's method of dealing with the integration of moment expressions with discontinuous first derivatives follows from the above discussion. The general rules to be followed when using this method are

(a) Take an origin at the left-hand end of the beam.

(b) Express the bending moment at a suitable section XX in the beam so as to include the effect of all the loads.

(c) Uniformly distributed loads must be made to extend to the right-hand end of the beam. It may be necessary to introduce negative compensating loads for this purpose.

(d) Put in square brackets all functions of length other than those involving x alone.

(e) Integrate as a whole any term in square brackets.

(f) When evaluating moment (curvature), slope or deflexion, neglect terms in square brackets when they become negative.

(g) In the moment equation, express concentrated moments in the form $M_1[x - a]^0$ where M_1 is the concentrated moment and $(x - a)$ is its point of application relative to section XX.

The following example illustrates all the rules mentioned above.

Example 5.6

Figure 5.8 shows a uniform beam of length $4L$ simply supported at the left-hand end A and at a distance L to the left of the right-

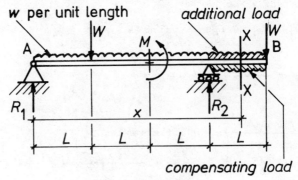

Figure 5.8

hand end, B. Concentrated loads W are applied at L to the right of A and at B. A concentrated anticlockwise moment M is applied at $2L$ to the right of A and a uniformly distributed load w per unit length extends from A to the right-hand support. Determine the slope and deflexion equations for the beam in terms of L, W, M, w and EI (the flexural rigidity).

If W is 4 kN, L is 4 m, w is 1 kN m^{-1} and M is 32 kN m, determine the position and magnitude of the maximum beam deflexion in terms of EI.

The section XX is taken between B and the right-hand support, at a distance x from A. The uniformly distributed load is extended to the right-hand end of the beam and a compensating load applied to preserve the original loading conditions.

Using the Macaulay rules, the moment at XX is

$$M_x = - R_1x + W[x - L] + M[x - 2L]^0$$
$$- R_2[x - 3L] + w\frac{x^2}{2} - \frac{w}{2}[x - 3L]^2$$

The load W at B has no moment about XX even when x takes its greatest value of $4L$, thus no term involving this load appears in the equation for M_x.

It will be seen that the additional and compensating loads to the right of XX are implied in the expression above since x can take any value between $3L$ and $4L$.

Integrating the moment expression we have

$$EI\frac{dy}{dx} = \frac{- R_1x^2}{2} + \frac{W}{2}[x - L]^2 + M[x - 2L]$$
$$- \frac{R_2}{2}[x - 3L]^2 + \frac{wx^3}{6} - \frac{w}{6}[x - 3L]^3 + A \tag{1}$$

and again

$$EIy = - R_1\frac{x^3}{6} + \frac{W}{6}[x - L]^3 + \frac{M}{2}[x - 2L]^2$$
$$- \frac{R_2}{6}[x - 3L]^3 + \frac{wx^4}{24} - \frac{w}{24}[x - 3L]^4 + Ax + B \tag{2}$$

To find the constants of integration we have the boundary conditions, $y = 0$ at $x = 0$ and $3L$. Thus

$$B = 0$$

and $A = \frac{3L^2}{2}R_1 - \frac{4WL^2}{9} - \frac{ML}{6} - \frac{9wL^3}{8}$

The slope and deflexion equations have now been obtained since it is only necessary to substitute for A, R_1 and R_2 in expressions 1 and 2 above.

Note that for vertical equilibrium

$$R_1 + R_2 = 2W + 3wL$$

131

For equilibrium of moments about A

$$WL + 4WL + 3wL \left(\frac{3L}{2}\right) = 3R_2L + M$$

thus $R_2 = \dfrac{5W}{3} + \dfrac{3wL}{2} - \dfrac{M}{3L}$

and $R_1 = \dfrac{W}{3} + \dfrac{3wL}{2} + \dfrac{M}{3L}$

If the numerical values are now substituted, we have

$$R_1 = R_2 = 10 \text{ kN}$$

and $A = + \dfrac{1064}{9} \text{ kN m}^2$

thus $EI \dfrac{dy}{dx} = -5x^2 + 2[x - 4]^2 + 32[x - 8] - 5[x - 12]^2$

$$+ \frac{x^3}{6} - \frac{1}{6}[x - 12]^3 + \frac{1064}{9} \tag{3}$$

and $EIy = -\dfrac{5}{3}x^3 + \dfrac{2}{3}[x - 4]^3 + 16[x - 8]^2$

$$- \frac{5}{3}[x - 12]^3 + \frac{x^4}{24} - \frac{1}{24}[x - 12]^4 + \frac{1064x}{9} \tag{4}$$

The usual procedure for determining the positions of maxima and minima in the deflexion curve is by setting the slope, dy/dx, equal to zero. In the slope equation (3), this operation is not straightforward since we do not know initially which of the square brackets should be retained and which should be ignored.

In order to establish the approximate zero-slope positions, slope values are determined at discrete points in the beam thus,

x (m)	0	4	8	12	16
$EI(dy/dx)$ (kN m^2)	+ 1064/9	+ 440/9	- 760/9	- 520/9	- 232/9

From the above table it is clear that a point of zero slope lies between 4 and 8 m from A. This point will have the maximum downward deflexion in the beam and is found from equation 3 by limiting the value of x to less than 8 m and setting the slope to zero, thus

$$3x_m{}^3 - 54x_m{}^2 - 228x_m + 2704 = 0$$

where x_m is the position of zero slope or maximum downward deflexion.

Solving this equation by trial in the range 4 m to 8 m we find that $x_m \simeq 5 \cdot 48$ m. After substituting for x_m in equation 4, the maximum downward deflexion is given by

$$y_{max} = + \frac{414}{EI}$$

From an examination of the table of slope values it can be seen that a larger upward deflexion could exist at the right-hand end of the beam. In order to check this, we put $x = 16$ m in equation 4 and find that

$$y_B = - \frac{146}{EI}$$

Thus it is confirmed that the numerically greater deflexion in the beam is downwards and at 5·48 m from the left-hand end A.

5.6 MOHR'S THEOREMS - THE MOMENT-AREA METHOD

An alternative method of calculating slopes and deflexions in beams can be derived from the basic curvature expression already given in equation 5.1. This may be re-written as

$$\frac{d^2y}{dx^2} = \frac{d\theta}{dx} = \frac{M}{EI} \qquad (5.2)$$

where $\theta = dy/dx$.

If the bending-moment diagram for a beam is determined, the curvature or M/EI diagram may be constructed by dividing moment ordinates by the flexural stiffness EI of the beam. Figure 5.9 shows the M/EI diagram for a beam AB under a general load system. Also shown in the figure is the so-called elastic line which represents the deflected shape of the beam.

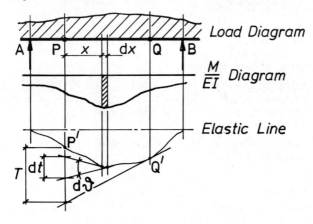

Figure 5.9

The area of a small element of the M/EI diagram (shown shaded in figure 5.9) is $M/EI \, dx$. From equation 5.2 this area is seen to be equal to the change in slope across the element or

$$d\theta = \frac{M}{EI} \, dx$$

Integrating this expression between limits P and Q in the beam we have

133

$$\theta_Q - \theta_P = \int_P^Q \frac{M}{EI} \, dx$$

= area of the M/EI diagram between P and Q.

From this result we obtain the first of Mohr's theorems

The change in slope between two points in a straight beam subjected to bending is equal to the area of the M/EI diagram between those points.

Compare this with the theorem in section 3.8.

If we now look at the elastic line in figure 5.9 we see that the tangents to the extremities of the element dx cut a vertical line through P at two points dt apart, where

$$dt = x \, d\theta = \frac{M}{EI} \, x \, dx$$

The total vertical interval T between P' on the elastic line and the intersection of the tangent drawn at Q' on the elastic line is obtained by integrating the above expression, thus

$$T = \int_P^Q \left(\frac{M}{EI} \, x \right) \, dx$$

The physical meaning of the integral is the first moment of area of the M/EI diagram between P and Q taken about P. This result leads to Mohr's second theorem

In a straight beam subjected to bending, the distance measured vertically from any point P' to the tangent drawn at a second point Q' is equal to the first moment of area of the M/EI diagram between P and Q, taken about P.

We shall now consider a number of simple examples to illustrate the moment-area method.

Example 5.7

Using the moment-area method, determine the slope and deflexion at the free end of the cantilever of example 5.2.

Referring to figure 5.10 and applying Mohr's first theorem, we have

$$\theta_B - \theta_A = \frac{WL}{EI} \times \frac{L}{2}$$

But $\theta_A = 0$, thus

$$\theta_B = \frac{WL^2}{2EI}$$

Applying Mohr's second theorem about A, we have

$$AA' = \left(\frac{WL}{EI} \times \frac{L}{2}\right) \frac{L}{3}$$

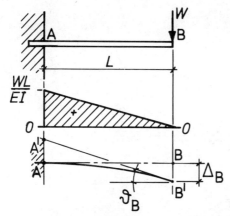

Figure 5.10

The distance AA' has no meaning as a deflexion but we note that

$$\Delta_B + AA' = L\theta_B$$

thus $\Delta_B = \dfrac{WL^2}{2EI} L - \dfrac{WL^3}{6EI} = \dfrac{WL^3}{3EI}$

A more direct way of obtaining this result is to apply the second theorem about B, then

$$BB' = \Delta_B = \left(\frac{WL}{EI} \times \frac{L}{2}\right) \frac{2L}{3} = \frac{WL^3}{3EI}$$

With the sign convention we have adopted for moments, displacements are positive if measured downwards with respect to the tangent.

Example 5.8

Determine the free-end slope and deflexion for the cantilever of example 5.3.

Refer to figure 5.11, then applying Mohr's first theorem to the M/EI diagram we have

$$\theta_B - \theta_A = \frac{wL^2}{2EI} \times \frac{L}{3}$$

since the area of a concave parabola is one-third of the enclosing rectangle (see appendix).

Since $\theta_A = 0$, we have

$$\theta_B = \frac{wL^3}{6EI} = \frac{WL^2}{6EI}$$

135

where $W = wL$.

Applying the second theorem about A and noting that the centroid of the M/EI diagram is at $L/4$ from A (see appendix), we have

$$AA' = \left(\frac{wL^2}{2EI} \times \frac{L}{3}\right) \frac{L}{4} = \frac{wL^4}{24EI}$$

but $AA' + \Delta_B = L\theta_B$

thus $\Delta_B = L \frac{wL^3}{6EI} - \frac{wL^4}{24EI} = \frac{wL^4}{8EI}$

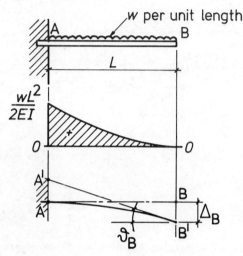

Figure 5.11

Alternatively, applying the second theorem about B, we have

$$\Delta_B = \left(\frac{wL^2}{2EI} \times \frac{L}{3}\right) \frac{3L}{4} = \frac{wL^4}{8EI} \text{ as before}$$

Example 5.9

Determine the end slope and the central deflexion in the beam of example 5.1.

Referring to figure 5.12, we apply the first moment-area theorem between A and the centre of the beam C, then

$$\theta_C - \theta_A = - \frac{wL^2}{8EI} \times \frac{2}{3} \left(\frac{L}{2}\right)$$

since the area of a convex parabola is two-thirds of the enclosing rectangle (see appendix).

But $\theta_C = 0$ by symmetry thus

$$\theta_A = \frac{wL^3}{24EI}$$

Applying the second theorem between A and C and about C we have

$$C'C'' = -\frac{wL^2}{8EI} \times \frac{2}{3}\left(\frac{L}{2}\right) \times \frac{3}{8}\frac{L}{2} = -\frac{wL^4}{128EI}$$

since the centroid of the parabola comprising the M/EI diagram between A and C is three-eighths of the base length AC from C (see appendix). The negative sign simply indicates that C' is above C".

But $CC'' = CC' + C'C'' = \frac{L}{2}\theta_A$

and $CC' = \Delta_C$

thus $\Delta_C = \frac{L}{2}\theta_A - C'C''$

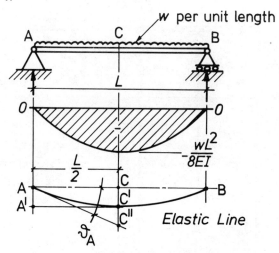

Figure 5.12

Substituting the numerical value of $C'C''$, we have

$$\Delta_C = \frac{L}{2} \times \frac{wL^3}{24EI} - \frac{wL^4}{128EI} = +\frac{5wL^4}{384EI}$$

If we apply the second theorem between A and C and about A, we have directly that

$$AA' = -\Delta_C = -\frac{wL^2}{8EI}\frac{2}{3}\left(\frac{L}{2}\right)\frac{5}{8}\left(\frac{L}{2}\right)$$

or $\Delta_C = +\frac{5wL^4}{384EI}$ as before

From these examples we see that it is not important which end of the M/EI diagram is used as the origin for determining the moment of the area provided the resulting displacement is properly interpreted.

The following is a more difficult problem. It can of course be solved by integration of the curvature expression but here Mohr's theorems are used in order to provide further illustration of the method.

Example 5.10

A uniform-section beam of length L is loaded by its own weight and is simply supported at two points. Find the distance between the two supports

(a) so that, with the supports at the same level, the two ends of the beam have zero slope

(b) so that the deviation from a straight line through the two supports is as small as possible.

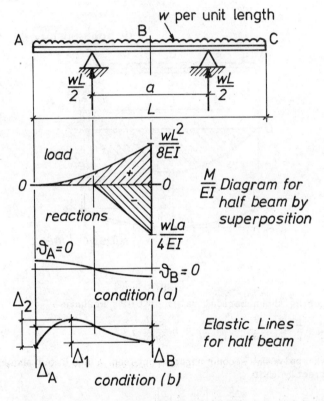

Figure 5.13

Let a be the distance between the supports and w the intensity of the uniformly distributed horizontal load then, referring to figure 5.13 we have for condition a that

$$\theta_A = \theta_B = \theta_C = 0$$

thus the area of the M/EI diagram between A and B must be zero, and

138

$$\frac{wL^2}{8EI}\left(\frac{1}{3} \times \frac{L}{2}\right) - \frac{wLa}{4EI}\left(\frac{1}{2} \times \frac{a}{2}\right) = 0$$

or $\frac{a}{L} = 0\cdot58$

For condition b we require Δ_1 or Δ_2 to be a minimum. This is so when $\Delta_1 = \Delta_2$, thus $\Delta_A = \Delta_B$ and the moment of area of the M/EI diagram between A and B about A must therefore be zero, hence

$$\frac{wL^2}{8EI}\left(\frac{1}{3} \times \frac{L}{2}\right)\left(\frac{3}{4} \times \frac{L}{2}\right) - \frac{wLa}{4EI}\left(\frac{1}{2} \times \frac{a}{2}\right)\left[\frac{L-a}{2} + \frac{2}{3} \times \frac{a}{2}\right] = 0$$

or $4\left(\frac{a}{L}\right)^3 - 12\left(\frac{a}{L}\right)^2 + 3 = 0$

Solution by trial gives

$\frac{a}{L} = 0\cdot55$

5.7 FORCE ANALYSIS OF STATICALLY INDETERMINATE BEAMS

If the application of the three equations of statical equilibrium is insufficient to determine all the forces and moments in a beam then that beam is said to be statically indeterminate. It is then necess-ary to investigate compatibility conditions to arrive at a solution.

The propped cantilever AB shown in figure 5.14 is one of the simplest examples of a statically indeterminate structure.

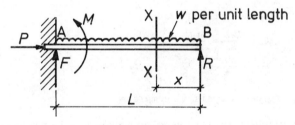

Figure 5.14

The equilibrium of the beam is ensured if

$P = 0$

$F + R = wL$

and $M + RL = \frac{wL^2}{2}$

Thus we have two unknown forces, F and R, and one unknown moment, M. There is one more unknown than can be obtained from the two available equations. The propped cantilever is therefore said to have one redundant force with respect to the requirements of statical equilibrium.

If we specify that the end B of the cantilever is to be propped

139

to the same height as the end A we have an additional condition with which to obtain a solution.

The problem could be solved by any of the methods that have been introduced so far. Here we give a solution employing direct integration of the moment expression. The reader is invited to attempt solutions based on the moment-area and superposition methods.

Referring to figure 5.14, the moment at XX is given by

$$M_x = EI \frac{d^2y}{dx^2} = - Rx + \frac{wx^2}{2}$$

thus $EI \frac{dy}{dx} = - \frac{Rx^2}{2} + \frac{wx^3}{6} + A$

and $EIy = \frac{Rx^3}{6} + \frac{wx^4}{24} + Ax + B$

at $x = L$, $dy/dx = 0$ and $y = 0$ thus

$$A = \frac{RL^2}{2} - \frac{wL^3}{6}$$

and $B = - \frac{RL^3}{3} + \frac{wL^4}{8}$

An additional boundary condition is that $y = 0$ when $x = 0$ thus

$$B = 0 \text{ or } R = \frac{3wL}{8}$$

From the equilibrium equations

$$F = wL - \frac{3wL}{8} = \frac{5wL}{8}$$
$$M = \frac{wL^2}{2} - \frac{3wL^2}{8} = \frac{wL^2}{8}$$

and the force analysis is complete.

Two examples of more complex problems will now be given.

Example 5.11

A uniform horizontal beam is firmly built-in at both ends to span a distance L. A frictionless hinge which transmits shear but no moment is inserted at midspan. The beam carries a vertical load W at a distance of $L/5$ from one end. Determine the values of the fixing moments at the ends and the shear force acting at the hinge. Sketch the bending-moment and shear-force diagrams for the beam.

The problem may be treated as two cantilevers as shown in figure 5.15. Since BC is a cantilever with a shear force Q acting at the free end, we have

$$\Delta_2 = \frac{Q(L/2)^3}{3EI} = \frac{QL^3}{24EI}$$

140

Applying Mohr's second theorem between A and B and about B we have

$$\Delta_1 = -\frac{QL}{2EI} \frac{L}{2} \frac{1}{2} \left(\frac{2}{3} \times \frac{L}{2}\right)$$

$$+ \frac{WL}{5EI} \frac{L}{5} \frac{1}{2} \left(\frac{2}{3} \times \frac{L}{5} + \frac{L}{2} - \frac{L}{5}\right)$$

or $$\Delta_1 = -\frac{QL^3}{24EI} + \frac{13WL^3}{1500EI}$$

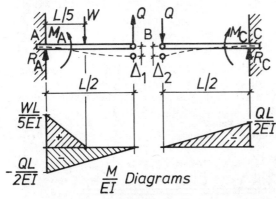

$\frac{M}{EI}$ Diagrams

Figure 5.15

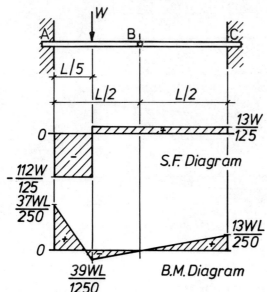

Figure 5.16

For continuity of displacement at the hinge, $\Delta_1 = \Delta_2$ thus

$$\frac{QL^3}{24EI} = -\frac{QL^3}{24EI} + \frac{13WL^3}{1500EI}$$

141

or $\quad Q = \dfrac{13W}{125}$

then $M_A = \dfrac{WL}{5} - \dfrac{QL}{2} = \dfrac{37WL}{250}$

and $\quad M_C = \dfrac{QL}{2} = \dfrac{13WL}{250}$

Figure 5.16 shows the shear-force and bending-moment diagrams for the problem.

Example 5.12

Figure 5.17 shows a beam 9 m long built-in at each end and carrying two loads each of 40 kN, symmetrically situated at 1·5 m on each side of midspan. The left-hand end of the beam sinks below the level of the right-hand end by 2·5 mm without tilting at either end. Find the fixing moments and reactions at the ends and sketch the shear-force and bending-moment diagrams for the beam. Take $EI = 25$ MN m^2.

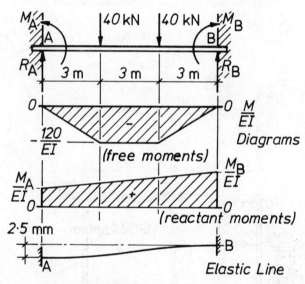

Figure 5.17

The moment-area method is used here. The reader is invited to obtain a solution by direct integration.

The total bending moment at any section in the beam is composed of the sum of the free bending-moment diagram (assuming the ends of the beam to be simply supported) and the bending-moment diagram due to the reactant moments M_A and M_B.

If the simple support reactions at A and B are R_A' and R_B' (both equal to 40 kN) and the reactions due to the presence of M_A and M_B alone are R_A'' and R_B'' we have

142

$$R_A = R_A' + R_A'' = 40 + \left(\frac{M_A - M_B}{9}\right) \text{ kN}$$

and $\quad R_B = R_B' + R_B'' = 40 + \left(\frac{M_B - M_A}{9}\right) \text{ kN}$

Since the slopes at A and B are both zero, the area of the combined M/EI diagrams is zero, then

$$-\frac{120 \times 6}{EI} + \frac{M_A}{EI} \times \frac{9}{2} + \frac{M_B}{EI} \times \frac{9}{2} = 0$$

or $\quad M_A + M_B = 160 \text{ kN m}$ $\qquad\qquad\qquad$ (1)

Applying Mohr's second theorem between A and B and about A we have

$$\frac{2 \cdot 5}{1000} = -\frac{720}{EI} \times \frac{9}{2} + \frac{M_A}{EI} \times \frac{9}{2} \times 3 + \frac{M_B}{EI} \times \frac{9}{2} \times 6$$

or $\quad M_A + 2M_B = 244 \cdot 63 \text{ kN m}$ $\qquad\qquad$ (2)

since $EI = 25 \times 10^3 \text{ kN m}^2$.

From equations 1 and 2 we have

$$M_A = + 75 \cdot 37 \text{ kN m}$$

and $\quad M_B = + 84 \cdot 63 \text{ kN m}$

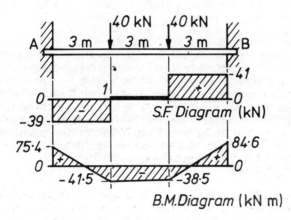

Figure 5.18

Taking moments about A, to obtain R_B we have

$$M_A - 3 \times 40 - 6 \times 40 - M_B + 9R_B = 0$$

thus $R_B = 41 \cdot 0 \text{ kN}$

but $R_A + R_B = 80$ kN

therefore $R_A = 39 \cdot 0$ kN.

Figure 5.18 shows the shear-force and bending-moment diagrams for the problem.

5.8 PROBLEMS FOR SOLUTION

1. A long straight bar of uniform rectangular cross-section and depth d is bent by two equal and opposite couples applied at its ends. It is found that the displacement at the midpoint of a chord length L is Δ. Show that the greatest longitudinal surface strain in the bar is given by $\varepsilon = 4d\Delta/L^2$.

2. A uniform-section beam of length L and flexural rigidity EI is simply supported at its ends and carries a single concentrated load W at a distance of $L/3$ from one end. Derive expressions for the deflexion (a) under the load, (b) at the centre, (c) at the point of maximum deflexion.
$(0 \cdot 0165, \ 0 \cdot 0178 \ \text{and} \ 0 \cdot 019 \times WL^3/EI)$

3. A steel beam of uniform section has a length of $6 \cdot 4$ m and is simply supported at points $4 \cdot 6$ m apart and $0 \cdot 90$ m from the ends. The beam carries three point loads, 20 kN at the left-hand end, 40 kN at the right-hand end and 120 kN at $3 \cdot 0$ m from the left-hand end. Determine the deflexion at each of the points of loading stating in each case whether the deflexion is upwards or downwards. Take $EI = 4$ MN m^2.
$(- \ 22 \cdot 8 \ \text{mm}, \ + \ 42 \cdot 4 \ \text{mm}, \ - \ 16 \cdot 45 \ \text{mm})$

4. A steel beam ABCD is 20 m long, (AB = CD = 5 m, BC = 10 m). The beam is simply supported at A and C. The loading consists of an anticlockwise moment of 10 kN m at A, a concentrated load of 50 kN together with a clockwise moment of 20 kN m at B, a clockwise moment of 30 kN m at C, a concentrated load of 10 kN at D and a uniformly distributed load of 5 kN m^{-1} running from B to C. Determine the required second moment of area and the minimum depth of the beam if the deflexion is not to be more than 1/400 of the length and the stress due to bending is not to exceed 150 MN m^{-2}. Neglect the self-weight of the beam and take $E = 200$ GN m^{-2}.
$(0 \cdot 44 \times 10^{-3} \ \text{m}^4, \ 572 \ \text{mm})$

5. A uniform beam ABCD has a span of 15 m and is simply supported at A and C. Concentrated vertical loads of 50 kN and 10 kN are placed at B and D respectively. If AB = BC = CD = 5 m and the value of EI for the beam is 25 MN m^2, determine the deflexion at D using the moment-area method.
(54 mm)

6. A uniform beam of length $5L$ is simply supported at its ends. A certain load system produces in the beam a constant shear-force W between the left-hand end and a point distant L along the beam. For

144

the next $2L$ the shear-force is zero and for the next $2L$ the shear
force varies linearly from zero to $-W$. Determine the deflexion of
a point distant $4L$ from the left-hand end of the beam.
$(1 \cdot 68WL^3/EI)$

7. A simply supported uniform beam AB of length L and flexural rigidity EI carries a distributed vertical load whose intensity varies
linearly from w per unit length at A to $2w$ per unit length at B. Determine the maximum deflexion and verify that this occurs at a point
distant approximately $0 \cdot 5L$ from A.
$(0 \cdot 0195wL^4/EI)$

8. An initially straight and horizontal cantilever of uniform section
and length L is rigidly built-in at one end and carries a uniformly
distributed load of intensity w/unit length for a distance $L/2$
measured from the built-in end. The second moment of area is I and
the modulus of elasticity is E. Determine, in terms of w, L, E and
I.

   (a) an expression for the slope of the cantilever at the end of
       the load

   (b) the deflexion at the free end of the cantilever

   (c) the force in a vertical prop that is to be applied at the
       free end in order to restore its level to that of the built-
       in end.

$$\left(\frac{wL^3}{48EI}, \; \frac{7wL^4}{384EI}, \; \frac{7wL}{128} \right)$$

9. A beam 8 m long carrying a uniformly distributed load of 50 kN/m
is lifted by three hydraulic jacks that are supplied by a common
pressure-line. One jack is under each end of the beam and one is
under the midpoint. The ram area of the middle jack is three times
that of the outer ones. Determine the level of the centre point of
the beam and the positions and magnitude of the maximum deflexion.
Deflexions are to be measured relative to the ends of the beam. Take
$EI = 25$ MN m^2.
$(4 \cdot 3$ mm, $2 \cdot 23$ and $5 \cdot 77$ m from one end, $5 \cdot 64$ mm$)$

10. A beam of uniform section is rigidly built-in at both ends. There
are concentrated loads of 40 kN at midspan and 20 kN at the left-hand
quarter-span point. A uniformly distributed load of 4 kN/m runs from
midspan to the right-hand support. Determine end moments and reactions.
$(88 \cdot 4$ and $82 \cdot 4$ kN m, $40 \cdot 6$ and $39 \cdot 4$ kN$)$

11. A beam of span $3 \cdot 75$ m is simply supported at its ends. It carries
a concentrated vertical load of 1 MN together with a concentrated
moment M at a point $1 \cdot 5$ m from the left-hand end. Determine the value
of M if the beam is to have zero slope under the load. Sketch the
shear-force and bending-moment diagrams for the beam and indicate
important values.
$(643$ kN m$)$

12. A uniform cantilever ABCD of length $4L$ is built-in at A and is simply supported at D to the same height as A. A concentrated vertical load W is carried at C distant $3L$ from A and an elastic support of stiffness k is provided at B distant $2L$ from A. If the reactions at B and D are to be equal show that

$$k = \frac{81EI}{28L^3}$$

where EI is the flexural rigidity of the cantilever.

6 STRAIN ENERGY

So far we have discussed the solution of problems in solid mechanics
by application of the conditions of equilibrium and compatibility
linked by the load-deformation characteristics of the material. In
this chapter we shall investigate an alternative approach using
energy methods.

6.1 THE BASIC ENERGY THEOREMS

A generalised force-deformation curve for a member made of a material
having a non-linear elastic stress-strain curve is shown in figure
6.1.

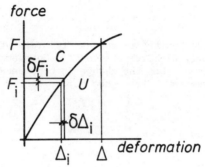

Figure 6.1

The area, U, under the curve represents the energy stored in the
member under a deflexion Δ. If we neglect losses of energy during
loading, this amount of energy must be equal to the work done on the
member as the force is gradually increased from zero to F. If we
again neglect losses, an equal amount of energy U is recoverable
during unloading.

U is defined as the strain energy in the member. The area C to
the left of the curve in figure 6.1 is defined as the complementary
strain energy. It has no clear physical interpretation except that

$$C = F\Delta - U \tag{a}$$

Suppose now that F_i and Δ_i represent the force and corresponding
deformation in a particular member of a structural system. If Δ_i
alone is increased by a small amount $\delta\Delta_i$, the change in strain energy
of the system is

$$dU = \frac{\partial U}{\partial \Delta_i} \; \delta\Delta_i \tag{b}$$

where the partial derivative represents the rate of change of strain
energy with respect to Δ_i. Work is done by the force F_i during this

deformation so that the change in strain energy is also given by reference to figure 6.1 as

$$dU = F_i \, \delta\Delta_i \qquad\qquad\qquad (c)$$

From equations b and c we have

$$F_i = \frac{\partial U}{\partial \Delta_i} \qquad\qquad\qquad (6.1)$$

This equation is a statement of what is usually referred to nowadays as Castigliano's first theorem.

If we now return to our particular member and increase its load by a small amount δF_i while all other loads in the system are held constant, then the change in the complementary strain energy of the system is

$$dC = \frac{\partial C}{\partial F_i} \, \delta F_i \qquad\qquad\qquad (d)$$

where the partial derivative represents the rate of change of complementary strain energy with respect to F_i. Complementary work is done during this process so that the change in complementary strain energy is also given by reference to figure 6.1 as

$$dC = \Delta_i \, \delta F_i \qquad\qquad\qquad (e)$$

From equations d and e

$$\Delta_i = \frac{\partial C}{\partial F_i} \qquad\qquad\qquad (6.2)$$

This equation is a statement of the complementary strain energy theorem usually attributed to Engesser.

Both the above theorems are perfectly general and apply equally well to non-linear or linear elastic materials. In the particular case of a linear elastic material we note that the strain energy U is equal to the complementary strain energy C so that equation 6.2 may be written as

$$\Delta_i = \frac{\partial U}{\partial F_i} \qquad\qquad\qquad (6.3)$$

This is a statement of what is usually referred to nowadays as Castigliano's second theorem.

Castigliano's first theorem requires that the strain energy of the system be written in terms of the deformations in order to perform the partial differentiation. The complementary strain energy theorem and Castigliano's second theorem on the other hand require that the energy be expressed as a function of the member forces. Both approaches are of value in analysis but in this chapter we shall confine our attention to the force method. We further limit our discussion to structural systems having a linear, elastic force-deformation

148

relationship so that attention will be directed to the application of Castigliano's second theorem, equation 6.3

Other energy theorems may be derived but these are outside the scope of this book.

6.2 EXPRESSIONS FOR STRAIN ENERGY

6.2.1 Strain Energy due to Axial Load

Consider the bar of length L and cross-sectional area A shown in figure 6.2a. An axial force F causes an axial deformation Δ. The material of the bar is linearly elastic thus the relationship (shown in figure 6.2b) between force and deformation is a straight line of slope EA/L where E is Young's modulus for the material.

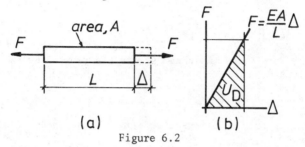

Figure 6.2

The work done by the force F is equal to the strain energy stored in the bar thus

$$U_D = \frac{1}{2} F\Delta \qquad\qquad (a)$$

where subscript D refers to direct stress. But

$$F = \frac{EA}{L} \Delta$$

thus $U_D = \dfrac{F^2 L}{2EA}$ \qquad\qquad (6.4)

In terms of stress, the strain energy per unit volume of bar is

$$u_D = \frac{\sigma^2}{2E} \qquad\qquad (6.5)$$

6.2.2 Strain Energy due to Shear

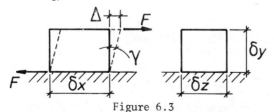

Figure 6.3

149

Consider the element of a linear elastic material of volume $\delta x \delta y \delta z$ shown in figure 6.3. A shear force F produces a deformation Δ as shown. Since the strain energy stored is equal to the work done, we have

$$U_S = \frac{1}{2} F\Delta \qquad\qquad (b)$$

where subscript s refers to shear; the shear stress τ in the element is given by

$$\tau = \frac{F}{\delta x \delta z} \qquad\qquad (c)$$

and for small deflexions the shear strain is

$$\gamma = \frac{\Delta}{\delta y} \qquad\qquad (d)$$

The shear stress and the shear strain are related by the shear modulus G, thus

$$\tau = \gamma G \qquad\qquad (e)$$

Substituting from equations c to e into equation b and eliminating γ we have

$$U_s = \frac{\tau^2}{2G} \delta x \delta y \delta z \qquad\qquad (f)$$

so that the shear strain energy per unit volume is

$$u_s = \frac{\tau^2}{2G} \qquad\qquad (6.6)$$

Expressions for shear strain energy in terms of forces or torques may be developed from equation 6.6 to suit particular cases.

6.2.3 Strain Energy due to Bending

Figure 6.4 shows a small element ABCD of an initially straight beam subjected to a moment M that is assumed constant along the element. The strain energy stored in the element is equal to the work done by the moment M, thus

$$dU_B = \frac{1}{2} M d\phi \qquad\qquad (g)$$

where subscript B refers to bending, $d\phi$ is the rotation of face BD with respect to face AC, and

$$d\phi = \frac{M}{EI} dx$$

If the beam is of length L we have for the whole beam

$$U_B = \int_0^L \frac{M^2}{2EI}\, dx$$

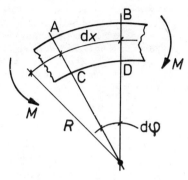

Figure 6.4

6.3 TORSION OF THIN-WALLED TUBES

As an example of a piece of analysis that requires the simple application of strain energy, we examine the torsion of thin-walled tubes.
This section is a continuation of the work of chapter 3.

The general analysis of the torsional behaviour of non circular
sections is a complex problem. It is possible, however, to derive
a simple approximate theory for such sections if they are closed
tubes having walls that are thin in comparison with their over-all
cross-sectional dimensions. This condition permits the assumption
that the shear stresses due to torsion are constant over the wall
thickness. Warping of cross-sections is assumed to be unrestrained
and the torque is constant along the tube.

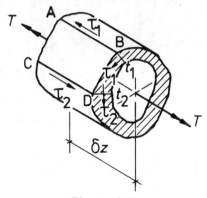

Figure 6.5

Consider the element of a thin-walled tube of arbitrary shape
shown in figure 6.5. The length of the element is δz. The wall

thickness may vary round the circumference but it is assumed that the thickness at a particular point does not vary along the length of the tube. Provided there are no axial restraints, a torque T will only produce shear stresses τ in the wall. Let the wall thickness and shear stress at point B in the wall be t_1 and τ_1 respectively. The corresponding values at D are t_2 and τ_2. Complementary shear stresses τ_1 and τ_2 therefore act on the longitudinal faces AB and CD of the rectangular section of wall ABCD. Since the wall thickness does not vary along the length of the tube, we have, for longitudinal equilibrium of ABCD.

$$\tau_1 t_1 \; \delta z = \tau_2 t_2 \; \delta z$$

or $\quad \tau_1 t_1 = \tau_2 t_2 = q$

Thus at a particular tube cross-section, the product of shear stress and wall thickness is constant at any point in the wall. This product is called the shear flow and is given the symbol q. The units of shear flow are those of force per unit length (i.e. $N \; mm^{-1}$ or $MN \; m^{-1}$ depending on the units chosen for τ and t). Note that since τt is constant the greatest shear stress occurs where the tube wall is thinnest.

6.3.1 The Bredt-Batho Equations

It is possible to determine the shear flow produced by a torque T if we consider the cross-section of a non-circular, thin-walled tube shown in figure 6.6. Suppose that the shear flow in the tube wall is q, then the reactant shear force acting on an element AB of length δs is δF, where $\delta F = q \; \delta s$.

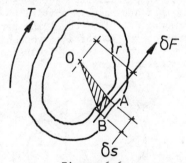

Figure 6.6

The torque about any point O produced by this shear force is

$$\delta T = r \; \delta F = qr \; \delta s$$

where r is the perpendicular distance from O to the δF vector.

Proceeding to the limit and integrating round the circumference of the tube, we have

$$T = q \int_s r \; ds \qquad\qquad\qquad (a)$$

152

The product r δs represents twice the area of the shaded triangle OAB in figure 6.6, thus the integral in equation a is equal to twice the area A_m contained within the middle line of the tube wall. Substituting for the integral, we have

$$T = 2qA_m \tag{6.8}$$

To determine the angle of twist of the tube we use the shear strain energy expression from section 6.2.2. The shear strain energy stored per unit length in the element AB is thus

$$\delta U_s = \left(\frac{q}{t}\right)^2 \frac{t \ ds}{2G} = \frac{q^2}{2G} \times \frac{ds}{t}$$

For a tube of length L, the total strain energy stored is

$$U_s = \frac{q^2 L}{2G} \int_s \frac{ds}{t}$$

where the integral is taken round the middle line of the tube.

If T is the constant applied torque and ϕ is the angle of twist over length L, the work done to store this amount of strain energy is $T\phi/2$, therefore

$$\frac{T\phi}{2} = \frac{q^2 L}{2G} \int_s \frac{ds}{t}$$

Substituting for the shear flow from equation 6.8 and rearranging we have

$$\phi = \frac{TL}{4GA_m^2} \int_s \frac{ds}{t} \tag{6.9}$$

Equations 6.8 and 6.9 are referred to as the Bredt-Batho equations for torsion of thin-walled tubes.

There is a similarity between equation 6.9 and equation 3.4 since if

$$C = \frac{4A_m^2}{\int_s \frac{ds}{t}}.$$

we have from equation 6.9

$$T = GC \frac{\phi}{L} \tag{6.10}$$

C is referred to as the torsion constant for the cross-section, it is only equal to the polar second moment of area, J, if the tube is circular in cross-section.

We can make use of equations 3.4 and 6.10 to evaluate the degree of approximation implicit in the Bredt-Batho equations. Consider a thin-walled circular section tube of wall thickness t and mean radius R. The angle of twist per unit length under a constant torque T is given by the exact theory (equation 3.4) as

$$\left(\frac{\phi}{L}\right)_{\text{exact}} = \frac{T}{GJ} \tag{a}$$

where

$$J = \frac{\pi}{2}\left[\left(R + \frac{t}{2}\right)^4 - \left(R - \frac{t}{2}\right)^4\right]$$

$$= 2\pi R^3 t\left[1 + \frac{1}{4}\left(\frac{t}{R}\right)^2\right]$$

The approximate theory (equation 6.10) gives

$$\left(\frac{\phi}{L}\right)_{\text{approx}} = \frac{T}{GC} \tag{b}$$

where

$$C = \frac{4A_m^2}{\displaystyle\int_S \frac{ds}{t}}$$

For a thin circular tube of constant thickness

$$\int_S \frac{ds}{t} = \frac{1}{t}\int_S ds = \frac{2\pi R}{t}$$

and $A_m = \pi R^2$

thus $C = 2\pi R^3 t$

From equations a and b

$$\left(\frac{\phi}{L}\right)_{\text{approx}} = \left(\frac{\phi}{L}\right)_{\text{exact}} \frac{J}{C}$$

or

$$\frac{(\phi/L)_{\text{approx}}}{(\phi/L)_{\text{exact}}} = \left[1 + \frac{1}{4}\left(\frac{t}{R}\right)^2\right]$$

The approximate equations therefore overestimate the angle of twist for a circular section tube. The error is less than 1% provided t is no greater than $R/5$.

The maximum shear stress calculated for the circular tube by the two methods are

$$\tau_{\text{approx}} = \frac{T}{2tA_m} = \frac{T}{2\pi R^2 t}$$

and $\tau_{exact} = \dfrac{T\left(R + \dfrac{t}{2}\right)}{J} = \dfrac{T\left[1 + \dfrac{1}{2}\left(\dfrac{t}{R}\right)\right]}{2\pi R^2 t\left[1 + \dfrac{1}{4}\left(\dfrac{t}{R}\right)\right]}$

therefore

$$\tau_{approx} = \tau_{exact}\left[\dfrac{1 + \left(\dfrac{t}{2R}\right)^2}{1 + \left(\dfrac{t}{2R}\right)}\right]$$

Thus the approximate equations underestimate the maximum shear stress in a thin circular tube. If t is equal to $R/5$, the error is about 8·2%. To reduce the error to less than 1%, t must not be greater than $R/50$.

It can be seen from this comparison that the approximate nature. of the Bredt-Batho equations is more apparent for stress calculation than for the determination of angle of twist.

Example 6.1

The thin-walled aluminium-alloy tube shown in figure 6.7 is subjected to a constant torque of 2 kN m. Determine the maximum shear stress in the tube and the angle of twist per metre length. G for the alloy is 28 GN m^{-2}.

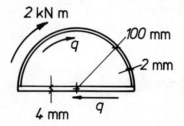

Figure 6.7

$A_m = \dfrac{\pi}{2}(100)^2 = 5000\pi \text{ mm}^2$

$\displaystyle\int_s \dfrac{ds}{t} = \dfrac{\pi \times 100}{2} + \dfrac{200}{4} = 50(1 + \pi)$

From equation 6.8 the shear flow is given by

$q = \dfrac{T}{2A_m} = \dfrac{2 \times 10^6}{2 \times 5000\pi} = 63\cdot7 \text{ N mm}^{-1}$

The maximum shear stress occurs where the tube wall is thinnest, thus

$\tau_{max} = \dfrac{63\cdot7}{2} \text{ N mm}^{-2} = 31\cdot8 \text{ N mm}^{-2}$

155

From equation 6.9, the angle of twist per unit length is given by

$$\frac{\phi}{L} = \frac{T}{4GA_m^2} \int_s \frac{ds}{t}$$

$$\frac{\phi}{L} = \frac{(2 \times 10^6)50(1 + \pi)}{4(28 \times 10^3)(5000\pi)^2} \text{ rad mm}^{-1}$$

or $\frac{\phi}{L} = 0 \cdot 015 \text{ rad m}^{-1} = 0 \cdot 86° \text{ m}^{-1}$

Example 6.2

Two coaxial thin-walled tubes of the same uniform thickness are made of the same material and are welded at their ends to rigid plates to form a composite tube. The inner tube is of circular cross-section, radius r, while the outer is of square cross-section of side a (where $a > 2r$).

A torque T is applied to the composite tube through the end plates Calculate the proportion of this torque that is carried by the inner tube and the ratio of the shear stresses in the two tubes.

Let T_i and T_o be the torques taken by the inner and outer tubes respectively, then

$$T = T_i + T_o \tag{1}$$

Since the angles of twist of the two tubes are identical, we have from equation 6.9

$$\frac{T_i L}{4GA_{m_i}^2} \int_{s_i} \frac{ds_i}{t} = \frac{T_o L}{4GA_{m_o}^2} \int_{s_o} \frac{ds_o}{t}$$

or $\frac{T_i}{(\pi r^2)^2} 2\pi r = \frac{T_o}{(a^2)^2} 4a$

thus $T_o = \frac{T_i a^3}{2\pi r^3}$ \hfill (2)

Substituting for T_o in equation 1 we obtain

$$T_i = \left(\frac{2\pi r^3}{2\pi r^3 + a^3}\right) T$$

From equation 6.8

$$T_i = 2\tau_i t A_{m_i}$$

and $T_o = 2\tau_o t A_{m_o}$

thus $\frac{\tau_i}{\tau_o} = \frac{T_i}{T_o} \times \frac{A_{m_o}}{A_{m_i}}$

or $\quad \dfrac{\tau_i}{\tau_o} = \left(\dfrac{2\pi r^3}{a^3}\right)\left(\dfrac{a^2}{\pi r^2}\right) = \dfrac{2r}{a}$

6.4 DIRECT APPLICATION OF STRAIN ENERGY

Certain simple problems involving structures carrying a single load and requiring the determination of a deflexion at and in the direction of the load may be solved by direct application of strain energy without recourse to Castigliano's second theorem.

6.4.1 Deflexion of Pin-jointed Frames

Once the forces in a pin-jointed frame have been determined by the methods of chapter 1 it is possible to write down from equation 6.4 the total strain energy stored as

$$U_D = \sum_{i=1}^{n}\left(\dfrac{F^2 L}{2EA}\right)_i \qquad\qquad\qquad (a)$$

where n is the number of members in the frame.

Since we are considering the case of a single applied load W and its corresponding deflexion Δ, the total work done on the frame by the external load must be given by

$$\text{work done} = \dfrac{1}{2} W\Delta \qquad\qquad\qquad (b)$$

The work done is equal to the strain energy stored so that from equations a and b

$$\Delta = \dfrac{1}{W}\sum_{i=1}^{n}\left(\dfrac{F^2 L}{AE}\right)_i \qquad\qquad\qquad (6.11)$$

We now apply this equation to determine the vertical deflexion Δ_V of the load point in the tripod of example 1.3. We shall take the cross-sectional area of all members as 1000 mm^2 and E as 200 GN m^{-2}, thus

$$\Delta_V = \dfrac{1}{100(10^{-3} \times 200 \times 10^6)}\left[(80\cdot0^2 + 8\cdot9^2)\sqrt{41} + (97\cdot5)^2\sqrt{77}\right]\text{ m}$$

or $\quad \Delta_V = 6\cdot2$ mm

6.4.2 Close-coiled Helical Springs

Figure 6.8 shows a section through a close-coiled helical spring of mean radius R consisting of n turns of circular-section wire of diameter d. The spring is subjected to an axial load W. If the coils are sufficiently close, it may be assumed that the only force action experienced by the wire is a torque WR. Let the over-all angle of twist in the wire caused by this torque be θ, then the strain energy stored in the spring is given by

157

$$U = \frac{1}{2} T\theta = \frac{WR\theta}{2} \qquad \text{(c)}$$

If the vertical deflexion of the load W is Δ, then the work done by the load is $W\Delta/2$. Equating the strain energy stored to the work done by the load we have

$$\Delta = R\theta \qquad \text{(d)}$$

but from equation 3.4 we have

$$\theta = \frac{TL}{GJ} \qquad \text{(e)}$$

where T is the torque on the spring wire ($= WR$), L is the length of the wire ($\approx 2\pi Rn$) and J is the polar second moment of area of the wire cross-section ($= \pi d^4/32$). Thus from equations d and e we have

$$\Delta = \frac{64WR^3 n}{Gd^4} \qquad (6.12)$$

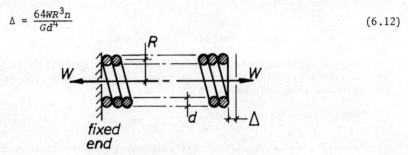

Figure 6.8

6.4.3 Deflexion of a Straight Beam

Consider the simply supported uniform beam with a central concentrated load shown in figure 6.9.

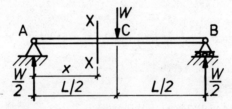

Figure 6.9

Let the vertical deflexion due to bending under the load be Δ_V, then the work done by W is $W\Delta/2$. The strain energy due to bending stored in the beam is the sum of the strain energies stored in lengths AC and CB which because of symmetry are equal, thus

$$U_B = 2\int_0^{L/2} \frac{1}{2EI} \left(\frac{Wx}{2}\right)^2 dx$$

or $\quad U_B = \dfrac{W^2 L^3}{96EI}$ $\qquad\qquad\qquad\qquad\qquad$ (f)

Equating strain energy to work done, we obtain

$$\dfrac{1}{2} W\Delta_v = \dfrac{W^2 L^3}{96EI}$$

or $\quad \Delta_v = \dfrac{WL^3}{48EI}$

Notice that since we have neglected the strain energy due to shear, the deflexion Δ_v will underestimate the true deflexion at this point. Shear deflexion is only significant in very short beams.

6.4.5 Thin Curved Members

The procedure discussed above may easily be extended to members curved in one plane. If the member is 'thin', deflexions due to bending predominate and strain energy due to shear and axial load may be neglected.

Consider the semi-circular beam of radius R shown in figure 6.10. One end is built-in and the other carries a horizontal load W.

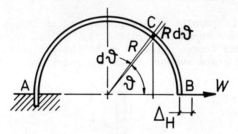

Figure 6.10

At a point C in the member displaced an angle θ from B, the bending moment M is given by

$$M = WR \sin \theta \qquad\qquad\qquad\qquad\qquad (g)$$

thus the total strain energy is

$$U_B = \int_0^\pi \dfrac{(WR \sin \theta)^2}{2EI} R \; d\theta$$

since $R \; d\theta = dx$. Thus

$$U_B = \dfrac{\pi}{4} \dfrac{W^2 R^3}{EI}$$

but the work done by the load is $W\Delta_H/2$, hence

$$\dfrac{1}{2} W\Delta_H = \dfrac{\pi}{4} \times \dfrac{W^2 R^3}{EI}$$

159

or $\quad \Delta_H = \frac{\pi}{2} \times \frac{WR}{EI}^3$ $\qquad$ (h)

6.5 DYNAMIC LOADING

If a body that possesses kinetic energy by virtue of its motion strikes an elastic member and is brought to rest, it can be conservatively assumed that all the kinetic energy is converted to strain energy in the member. Equally, if the body falls from a certain height, the potential energy that the body has lost will be converted into strain energy in the member.

Suppose a load W falls a height h on to an elastic member and produces a maximum deflexion δ at the point of impact. The potential energy, V, transferred to the member is given by

$$V = U = W (h + \delta) \qquad (a)$$

Let W_e be the equivalent static load to produce the same deflexion in the member, then the work done by W_e is $W_e\delta/2$. But the work done is equal to the strain energy stored thus

$$\frac{1}{2} W_e\delta = W(h + \delta) \qquad (b)$$

Consider the problem of a vertical rod of length L and cross-sectional area A shown in figure 6.11. A load W, concentric with the rod, is dropped from a height h on to the end stop. The rod is suspended from a rigid support, thus the reaction at this point does no work.

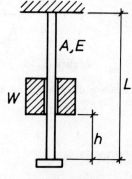

Figure 6.11

For a rod in tension

$$\delta = \frac{W_e L}{AE}$$

After substituting in equation b we obtain

$$\sigma^2 - 2\sigma_{st}\sigma - \frac{2Eh}{L} \sigma_{st} = 0 \qquad (c)$$

160

where σ, the dynamic stress = W_e/A and σ_{st}, the static stress = W/A.

From equation c, the maximum dynamic stress is

$$\sigma_{max} = \sigma_{st} \left[1 + \sqrt{\left(1 + \frac{2Eh}{L\sigma_{st}}\right)} \right]$$

Note that even when h is zero the maximum dynamic stress is twice the static stress.

Example 6.3

A small light piston with a cross-sectional area of 130 mm^2 compresses oil in a rigid container. When a body having a weight of 45 N is gradually applied to the piston its movement is found to be 16 mm. If a body having a weight of 18 N now falls from a height of 80 mm on to the 45 N weight, determine the magnitude of the pressure pulse. Neglect all energy losses.

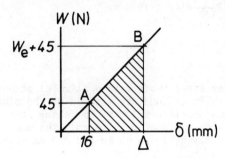

Figure 6.12

Figure 6.12 shows the load-deflexion characteristic for the system. W_e is the static equivalent of the 18 N weight and Δ is the final maximum deflexion of the piston.

From the figure, the static stiffness of the system is given by

$$\frac{W}{\delta} = \frac{45}{16} = \frac{(W_e + 45)}{\Delta} \text{ N mm}^{-1} \tag{1}$$

The potential energy lost by the two weights in moving from A to B on the characteristic is given by

$$V = 18(80 + \Delta - 16) + 45(\Delta - 16) \text{ N mm}$$

The strain energy stored in the system is the shaded area in figure 6.12 or

$$U = \frac{1}{2}(W_e + 45)\Delta - \frac{1}{2}(45)(16) \text{ N mm}$$

If there are no losses we may equate potential energy lost to

161

strain energy stored, then substituting for $(W_e + 45)$ from equation 1 we obtain the following quadratic in Δ

$$\Delta^2 - 44 \cdot 8 \Delta - 563 \cdot 2 = 0$$

from which the positive root gives

$$\Delta = 55 \text{ mm}$$

From equation 1

$$(W_e + 45) = \frac{45}{16} \ (55) = 154 \cdot 7 \text{ N}$$

or $\quad W_e = 109 \cdot 7 \text{ N}$

The maximum pressure in the oil is therefore given by

$$p_{max} = \frac{(W_e + 45)}{130} = 1 \cdot 19 \text{ N mm}^{-2}$$

Example 6.4

A solid circular steel shaft with a flywheel at one end rotates at 2 rev/s. The shaft is suddenly stopped at the other end. Determine the maximum possible value of the shear stress if the shaft is 1·5 m long with a diameter of 50 mm. The mass and radius of gyration of the flywheel are 45 kg and 250 mm respectively. Take G as 80 GN m^{-2}.

The kinetic energy of the flywheel is given by $I\omega^2/2$, where the mass moment of inertia is

$$I = mk^2 = (45)(0 \cdot 25)^2 = 2 \cdot 8 \text{ kg m}^2$$

and the angular velocity

$$\omega = 2\pi(2) = 4\pi \text{ s}^{-1}$$

The total kinetic energy lost by the flywheel is thus

$$\tfrac{1}{2}(2 \cdot 8)(4\pi)^2 = 222 \cdot 0 \text{ kg m}^2 \text{ s}^{-2} = 222 \cdot 0 \text{ N m}$$

since $1 \text{ N} = 1 \text{ kg m s}^{-2}$.

The strain energy stored in a shaft subject to a torque T producing an angle of twist ϕ is given by

$$U_T = \tfrac{1}{2} \ T\phi$$

From equation 3.6 we have

162

$$T = \tau_{max} \frac{J}{R}$$

and $\quad \phi = \tau_{max} \frac{L}{GR}$

thus $\quad U_T = \tau_{max}^2 \frac{JL}{2GR^2}$

but $\quad J = \frac{\pi R^2}{2}$

therefore

$$U_T = \frac{\tau_{max}^2}{4G} \pi R^2 L$$

Since the strain energy gained by the shaft is equal to the kinetic energy lost by the flywheel we have

$$U_T = 222 \cdot 0 \ N \ m$$

or $\quad \tau_{max} = \sqrt{\left[\frac{4G(222 \cdot 0)}{\pi R^2 L} \right]} \ N \ m^{-2}$

substituting the values for G, R and L we obtain

$$\tau_{max} = 155 \cdot 3 \ MN \ m^{-2}$$

6.6 APPLICATIONS OF CASTIGLIANO'S SECOND THEOREM TO THE DETERMINATION OF DEFLEXIONS

6.6.1 Deflexions in Pin-jointed Frames

In section 6.4.1 we determined the deflexion in a pin-jointed frame carrying a single load. The direct use of strain energy was limited to finding the deflexion at the load point in the direction of the line of action of the load. The use of Castigliano's second theorem will allow us to determine deflexions at any point in a multi-loaded frame.

Suppose we wish to determine the deflexion at a particular point in a frame. We place a load P at the point acting in the direction of the desired deflexion, Δ. From equation 6.3

$$\Delta = \frac{\partial U_D}{\partial P} \qquad\qquad\qquad\qquad (a)$$

where U_D is the total strain energy in the frame and from equation a of section 6.4.1 is given by

$$U_D = \sum_{i=1}^{n} \left(\frac{F^2 L}{2EA} \right)_i \qquad\qquad\qquad\qquad (b)$$

163

Applying equation a above, we have

$$\Delta = \sum_{i=1}^{n} \left(\frac{\partial F}{\partial P} \times \frac{FL}{EA} \right)_i \qquad \text{(c)}$$

where the partial differential coefficient is interpreted physically as the ratio of the force in a member to the load producing it. For convenience we allow P to be a unit load; this makes no difference to $\partial F/\partial P$ since it is a constant ratio for all ordinary frames, then

$$\Delta = \sum_{i=1}^{n} \left(\frac{KFL}{EA} \right)_i \qquad \text{(d)}$$

where K is a dimensionless number equivalent to the force in a member produced by a force of unit magnitude (no units) acting at, and in the direction of, the deflexion Δ. The following example will illustrate the use of equation d.

Example 6.5

Determine the x-ward deflexion of the load point in the tripod of example 1.3. EA is 200 MN.

The member forces, F, due to the 100 kN vertical load have already been determined. To obtain the values of K we place a unit horizontal load (acting in the x-direction) at O in figure 1.8. Using tension coefficients the values of K shown in the table below are calculated. The negative sign denotes compression.

Member	F (kN)	K	L (m)	KFL (kN m)
OA	+ 80·0	+ 0·8	$\sqrt{41}$	+ 409·8
OB	- 8·9	- 0·8	$\sqrt{41}$	+ 45·6
OC	- 97·5	0	$\sqrt{77}$	0

$$\sum KFL = + 455\cdot5 \text{ kN m}$$

Since EA is a constant for all members we have from equation d

$$\Delta_H = + \frac{455\cdot4}{200 \times 10^3} \text{ m} = + 2\cdot3 \text{ mm}$$

The positive sign indicates that the deflexion is in the direction assumed for the unit load, namely in the positive x-direction of figure 1.8. The reader is invited to confirm that the deflexion in the z-direction is -2·8 mm.

6.6.2 Deflexions in Thin Curved Members

In a similar manner to the above treatment for deflexions in pin-jointed frames we may determine the deflexion Δ at a particular point in a thin curved member. We place a load P at the point

164

acting in the direction of the required deflexion Δ. If we only consider strain energy due to bending we have, from equations 6.3 and 6.7 that

$$\Delta = \frac{\partial U_B}{\partial P} = \int \frac{\partial M}{\partial P} \times \frac{M}{EI} \ dx \qquad (a)$$

The physical interpretation of $\partial M/\partial P$ is the moment produced in the member by the load P divided by P, in other words, it represents a moment arm and thus has the dimensions of length. Since the moment arm is independent of the value of P we may replace P by a unit load.

In order to illustrate the method we shall determine the vertical deflexion of the load point B in the semi-circular beam shown in figure 6.10.

If a unit load acting vertically upwards is placed at B the corresponding moment arm with respect to C is given by

$$\frac{\partial M}{\partial P} = R(1 - \cos \theta)$$

but $M = WR \sin \theta$ and $dx = R \ d\theta$, thus

$$\Delta_v = \frac{WR^3}{EI} \int_0^\pi \sin \theta (1 - \cos \theta) d\theta$$

or $\quad \Delta_v = \dfrac{2WR^3}{EI}$

The deflexion is positive thus the load point moves up.

Example 6.6

The davit shown in figure 6.13 is subject to a vertical load W at the free end C. Determine the height of the straight portion AB in

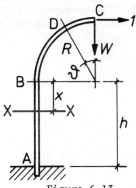

Figure 6.13

terms of the radius of BC if the horizontal and vertical components of the deflexion at C are to be equal. The davit has a constant value of EI throughout and strain energy due to bending only is to be considered.

The davit consists of two parts, the quadrant BC and the straight column AB, thus two separate strain-energy expressions are required. The direct application of strain energy could be used to determine the vertical deflexion at C, but to illustrate the method the unit load approach will be used here.

For the quadrant BC, the moment at D is given by

$$M_{BC} = WR \sin \theta$$

The moment arm with respect to D due to a vertical unit load at C is thus

$$\left(\frac{\partial M}{\partial P}\right)_{BC} = R \sin \theta$$

For the column AB, the moment at section XX is

$$M_{AB} = WR$$

and

$$\left(\frac{\partial M}{\partial P}\right)_{AB} = R$$

thus the vertical component of deflexion at C is given by

$$\Delta_V = \int_0^{\pi/2} R \sin \theta \; \frac{WR \sin \theta}{EI} R \; d\theta + \int_0^h R \frac{WR}{EI} \; dx$$

hence

$$\Delta_V = \frac{WR^3}{EI} \left(\frac{\pi}{4} + \frac{h}{R}\right) \tag{1}$$

To determine the horizontal component of the deflexion at C, we apply a unit horizontal load. The values of M_{BC} and M_{AB} are as before but now

$$\left(\frac{\partial M}{\partial P}\right)_{BC} = R(1 - \cos \theta)$$

and

$$\left(\frac{\partial M}{\partial P}\right)_{AB} = (R + x)$$

thus

$$\Delta_H = \int_0^{\pi/2} R(1 - \cos \theta) \; \frac{WR \sin \theta}{EI} R \; d\theta + \int_0^h (R + x) \frac{WR}{EI} \; dx$$

hence

$$\Delta_H = \frac{WR^3}{2EI} \left(1 + \frac{h}{R}\right)^2 \tag{2}$$

166

Since the two components of deflexion are to be equal we have from equations 1 and 2 that

$$\frac{h}{R} = \sqrt{\left(\frac{\pi}{2} - 1\right)} = 0 \cdot 755$$

6.7 BEAMS CURVED IN PLAN

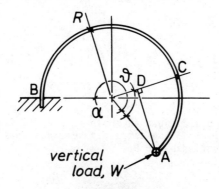

Figure 6.14

Figure 6.14 shows the plan view of a circular-section horizontal beam AB bent into a circular arc of radius R and subjected to a concentrated vertical load W at A.

The force action at any point C in the beam consists of a bending moment M, a torque T and a shear force which may usually be neglected. The moment and torque are given by

$$M = W \times AD$$

and $\quad T = W \times CD$

In problems of this type it is necessary to take account of strain energy due to bending and torsion. The strain energy due to torsion in an element of the beam of length $R\,d\theta$ is given by

$$dU_T = \frac{1}{2} T \, d\phi = \frac{T^2}{2GJ} R \, d\theta$$

thus

$$U_T = \int \frac{T^2 R}{2GJ} \, d\theta \qquad\qquad (a)$$

For the problem shown in figure 6.14 we have

$$AD = R \sin \theta$$

and $\quad CD = R(1 - \cos \theta)$

thus the strain energy in the beam is

167

$$U = U_B + U_T$$

or $\quad U = \int_0^\alpha \frac{(WR \sin \theta)^2}{2EI} \; R \; d\theta + \int_0^\alpha \frac{[WR(1 - \cos \theta)]^2}{2GJ} \; R \; d\theta \qquad$ (b)

If the vertical deflexion of the load point is required we may equate the work done by the load to the strain energy stored thus

$$\frac{1}{2} W\Delta_V = U$$

or $\quad \Delta_V = \frac{WR^3}{EI} \int_0^\alpha \sin^2 \theta \; d\theta + \frac{WR^3}{GJ} \int_0^\alpha (1 - \cos \theta)^2 d\theta \qquad$ (c)

Example 6.7

The circular-section beam AB shown in plan in figure 6.15 lies in the horizontal plane and carries a vertical load W at B. Determine the vertical deflexion of B.

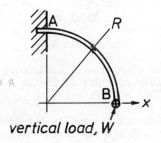

Figure 6.15

The vertical deflexion may be obtained by applying equation c above, then

$$\Delta_V = \frac{WR^3}{EI} \int_0^{\pi/2} \sin^2 \theta \; d\theta + \frac{WR^3}{GJ} \int_0^{\pi/2} (1 - \cos \theta)^2 \; d\theta$$

or $\quad \Delta_V = \frac{WR^3}{4EI} \left[\pi + (3\pi - 8) \frac{EI}{GJ} \right]$

6.8   PROBLEMS FOR SOLUTION

1. A steel bar of rectangular section 40 mm wide and 10 mm deep is arranged in a horizontal position as a beam with fixed ends and span 1 m. A load of 120 N is allowed to fall freely onto the beam at mid-span. Find the height above the initial position of the beam from which the load must fall in order to produce in the beam a maximum stress of 120 MN m^{-2}. E = 200 GN m^{-2}.
(8·34 mm)

2. A steel rod 12 mm diameter and 2·5 m long hangs vertically from a close-coiled helical spring made of 16 mm diameter rod and having 6 effective turns of 100 mm mean diameter. The top of the spring is

168

rigidly attached and the spring and rod are coaxial. A load of 2·5
kN slides on the rod and strikes a stop at the end of the rod after
falling freely through a distance of 15 mm.

Determine

(a) the maximum extensions of spring and rod
(b) the maximum shear stress in the spring and tensile stress in
the rod.

Take E = 200 GN m^{-2}, G = 80 GN m^{-2}.
(57·5 and 0·625 mm, 391 and 55·5 MN m^{-2})

3. Three high-strength cables (1, 2 and 3) of equal cross-sectional
area are suspended from a rigid ceiling and support a vertical load
of 80 kN. Figure 6.16 shows two views of the cable-suspension system.

Determine the largest cable stress if the vertical deflexion of
the point of application of the load is not to exceed 10 mm. Take
E = 200 GN m^{-2}.
(298 MN m^{-2})

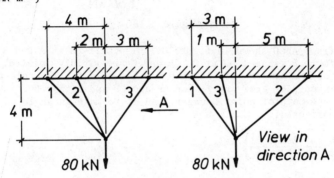

Figure 6.16

4. The davit shown in figure 6.17 is built-in at the foot and is made
of solid circular-section steel bar of diameter 100 mm. What is the

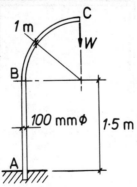

Figure 6.17

169

greatest value that the load W can have if the maximum bending stress
in the davit is not to exceed 160 MN m^{-2} and if the vertical deflexion
at C is not to exceed 30 mm? When determining the deflexion, take
account of strain energy due to bending only. For steel E = 200 GN
m^{-2}.

(12·9 kN)

5. Determine the vertical and horizontal deflexions of point A in the
pin-jointed frame shown in figure 6.18. All members have EA = 20 MN,
AB = AC = BC = 1 m, DB = 0·5 m.

(9·2 mm, 1·45 mm)

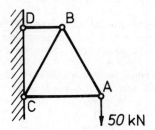

Figure 6.18

6. The frame ABCD shown in figure 6.19 is built-in at A and a horizontal force H is applied at D. B and C are rigidly jointed. The
lengths of the members are AB = CD = h and BC = L. The relevant
second moments of area are I_1 for AB and CD and I_2 for BC. Show
that if the path of point D under load is to make an angle of 45°
to the horizontal.

$$\frac{I_1}{I_2} = \frac{h}{3L}\left(\frac{4h - 3L}{L - 2h}\right)$$

Ignore the effects of axial load and shear force on the frame.

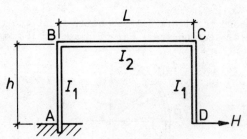

Figure 6.19

7. Figure 6.20 shows the plan view of a horizontal beam subjected to
a vertical load P at D. Determine the vertical deflexion of the beam
at the load point.

$$\left[2Pa^3\left(\frac{5}{3EI} + \frac{3}{GJ}\right)\right]$$

170

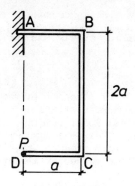

Figure 6.20

8. A cantilever carrying a distributed load of intensity w per unit length is semi-circular in plan. If the cantilever has a circular cross-section and is made of steel for which $E = 2 \cdot 5G$, show that the vertical deflexion of the free end is approximately

$$\frac{8 \cdot 2wR^4}{EI}$$

9. A thin-walled rectangular box beam of length 3 m is built-in at both ends and supports concentrated torques of 250 N m at 1 m from each end, together with a uniformly distributed torque of 2 N m mm^{-1} over the middle third of the span. All the applied torques act in the same sense. The long walls of the box are 400 mm in length and have a thickness of 1·5 mm while the short walls are 200 mm long with a thickness of 2 mm. Determine the maximum shear stress in the box and the maximum angle of twist relative to the ends. Take $G = 81$ GN m^{-2}. Neglect warping.
(5·21 N mm^{-2}, 0·53 × 10^{-3} rad)

10. A hexagonal cross-section tube is made up from six aluminium-alloy plates 200 mm wide and 10 mm thick. Determine the maximum shear stress and the angle of twist per unit length under a constant torque of 60 kN m. $G = 30$ GN m^{-2}.
(14·4 N mm^{-2}, 0·08° m^{-1})

7 BIAXIAL STRESS AND STRAIN

Previously the evaluation of stresses in members has been relatively simple since we have considered stress due to a single force action (direct stress due to axial load, bending stress due to moment, or shear stress due to torque) or to a combination of force actions (axial load and moment) which produce stresses acting uniaxially, the resultant of which may be obtained by algebraic addition.

Certain combinations of forces (such as axial load and torque) produce stresses acting in planes at right angles to each other. The resultants of such a biaxial stress state cannot be obtained by simple addition.

The use of the term biaxial implies a two-dimensional analysis for stresses in the xy-plane. The stress normal to this plane is assumed to be zero, a condition referred to as plane stress.

A later part of the chapter is devoted to the analysis of the bi-axial state of strain, the assumption being that strain in the direction normal to the xy-plane is zero. This condition is referred to as plane strain. Since strains are possible without stresses in the same direction it will be appreciated that plane stress and plane strain are not necessarily equivalent.

7.1 MOHR'S CIRCLE FOR STRESS

Figure 7.1 shows a small element of material that is subjected to direct stresses σ_x and σ_y together with complementary shear stresses τ.

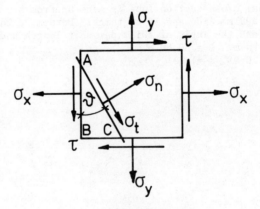

Figure 7.1

We wish to determine the stresses σ_n and σ_t acting on a plane AC inclined at an angle θ to the plane on which the stress σ_x acts. We

172

assume that the element is of unit thickness and that AC is of unit length, thus

AB = cos θ and BC = sin θ

To satisfy the conditions of statical equilibrium for ABC we resolve the *forces* due to the stresses in the directions normal to and along AC, thus eventually we obtain

$$\sigma_n = \frac{1}{2}(\sigma_x + \sigma_y) + \frac{1}{2}(\sigma_x - \sigma_y)\cos 2\theta + \tau \sin 2\theta \qquad (7.1)$$

and $\sigma_t = \frac{1}{2}(\sigma_x - \sigma_y)\sin 2\theta - \tau \cos 2\theta$ (7.2)

If θ is eliminated between equations 7.1 and 7.2 we obtain

$$[\sigma_n - \frac{1}{2}(\sigma_x + \sigma_y)]^2 + \sigma_t^2 = [\frac{1}{2}(\sigma_x - \sigma_y)]^2 + \tau^2 \qquad (a)$$

Equation a is the equation of a circle on axes σ_n and σ_t whose centre is at

$$[\frac{1}{2}(\sigma_x + \sigma_y),\ 0]$$

and whose radius is

$$\sqrt{\{[\frac{1}{2}(\sigma_x - \sigma_y)]^2 + \tau^2\}}$$

Figure 7.2 shows the circle diagram drawn from equation a. This diagram is known as Mohr's circle for stress. Compare with the circle of second moments of area in chapter 4.

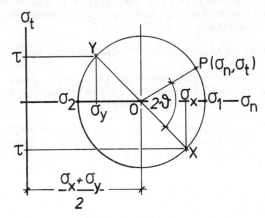

Figure 7.2

A radius vector OP at an angle 2θ measured anticlockwise from OX intersects the circle at P. The coordinates of P are then the normal and tangential (shear) stresses on the plane AC inclined at an angle θ to the plane on which the stress σ_x acts.

The maximum and minimum numerical values of σ_n are σ_1 and σ_2 respectively. They occur on planes 90° apart. The stresses σ_1 and σ_2 are known as principal stresses and the planes on which they act are principal planes. It is clear from figure 7.2 that the shear stress is zero on the principal planes.

If angle θ_1 defines the position of the principal planes, we have from equation 7.2 with $\sigma_t = 0$ that

$$\theta_1 = \frac{1}{2}\tan^{-1}\left(\frac{2\tau}{\sigma_x - \sigma_y}\right) \tag{b}$$

Substituting for θ_1 in equation 7.1 or setting $\sigma_t = 0$ in a we obtain the values of the principal stresses, thus

$$\sigma_{1,2} = \frac{1}{2}(\sigma_x + \sigma_y) \pm \{[\frac{1}{2}(\sigma_x - \sigma_y)]^2 + \tau^2\}^{1/2} \tag{c}$$

The maximum value of the shear stress is equal to the radius of Mohr's circle or half the difference of the principal stresses. A direct stress equal to the mean value of σ_x and σ_y is associated with the maximum shear stress. Planes of maximum shear stress are 90° apart and bisect the angle between the principal planes.

The special case when σ_x and σ_y are principal stresses and are equal and opposite in sign produces a state of pure shear on planes bisecting the principal planes. The centre of the stress circle is then at the origin and the numerical values of the principal stresses and the maximum shear stress are all equal. For pure shear, the plane stress and plane strain conditions are equivalent.

Mohr's stress circle is a semi-graphical aid to solving the bi-axial stress equations 7.1 and 7.2. These equations could be solved directly if desired.

7.2 MAXIMUM SHEAR STRESS UNDER PLANE STRESS CONDITIONS

The maximum shear stress shown on the Mohr's circle diagram in figure 7.2 is the maximum shear stress in the xy-plane. This may not be the maximum shear stress in the element.

Suppose for a moment that we consider a triaxial state of stress with principal stresses σ_1, σ_2 and σ_3 acting mutually at right angles. If $\sigma_1 > \sigma_2 > \sigma_3$, the maximum shear stresses in each plane are as follows

$$\tau_{max} = \frac{1}{2}(\sigma_1 - \sigma_2) \text{ in the 1,2 plane}$$

$$\tau_{max} = \frac{1}{2}(\sigma_1 - \sigma_3) \text{ in the 1,3 plane}$$

and $\tau_{max} = \frac{1}{2}(\sigma_2 - \sigma_3)$ in the 2,3 plane

For the plane stress condition, $\sigma_3 = 0$ thus the largest of the maximum shear stresses lies in the 1,3 plane and has the value

$$\tau_{max} = \frac{1}{2} \sigma_1$$

For plane stress with σ_2 negative and σ_1 positive, the largest maximum shear stress is in the xy- (1,2) plane. The two cases are shown on the Mohr's circle diagrams in figure 7.3.

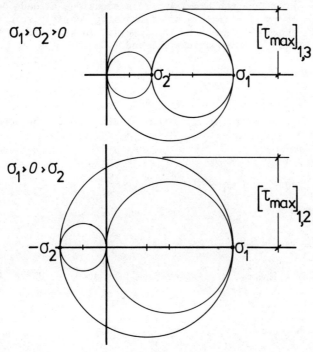

$\sigma_1 > \sigma_2 > 0$

$[\tau_{max}]_{1,3}$

$\sigma_1 > 0 > \sigma_2$

$[\tau_{max}]_{1,2}$

Figure 7.3

Example 7.1

Direct stresses of 80 MN m^{-2} tension and 60 MN m^{-2} compression are applied to an elastic material at a certain point on planes at right

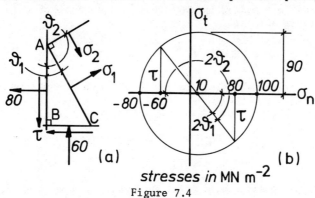

(a)

(b)

stresses *in* MN m^{-2}

Figure 7.4

175

angles to one another. If the greater principal stress in the material is limited to 100 MN m^{-2}, to what shear stress may the material be subjected on the given planes and what will then be the maximum shear stress at the point?

This example can, of course, be solved from first principles. Here, however, we use the biaxial stress equations already developed and then go on to illustrate a solution using Mohr's circle diagram. Figure 7.4a shows an element of the material in which AC is a principal plane. From equations 7.1 and 7.2

$$\sigma_1 = 10 + 70 \cos 2\theta_1 + \tau \sin 2\theta_1 \tag{1}$$

and $\quad 0 = 70 \sin 2\theta_1 - \tau \cos 2\theta_1 \tag{2}$

but σ_1 is limited to 100 MN m^{-2}, thus eliminating $2\theta_1$ between equations 1 and 2 we obtain

$$\sqrt{(70^2 + \tau^2)} = 90 \text{ MN m}^{-2}$$

or $\quad \tau = \sqrt{3200} = 56 \cdot 6 \text{ MN m}^{-2}$

From equation 2

$$\theta_1 = \frac{1}{2} \tan^{-1}\left(\frac{\tau}{70}\right) = 19°28'$$

The other principal plane therefore lies at $\theta_2 = 109°28'$ hence

$$\sigma_2 = 10 + 70 \cos 2\theta_2 + \tau \sin 2\theta_2 \text{ MN m}^{-2}$$

or $\quad \sigma_2 = -80 \text{ MN m}^{-2}$

The maximum shear stress is in the plane containing the principal stresses σ_1 and σ_2 and is given by

$$\tau_{max} = \frac{1}{2} (\sigma_1 - \sigma_2) = 90 \text{ MN m}^{-2}$$

The Mohr's circle for the example is shown in figure 7.4b. Since σ_x is + 80 MN m^{-2} and σ_y is - 60 MN m^{-2}, the centre of the circle lies on the σ_n axis at

$$\frac{1}{2}(\sigma_x + \sigma_y) = + 10 \text{ MN m}^{-2}$$

the maximum principal stress is 100 MN m^{-2} thus the maximum shear stress which is equal to the radius of the circle is (100 - 10) = 90 MN m^{-2}. The shear stress τ is then given by

$$\tau = \sqrt{(90^2 - 70^2)} = 56 \cdot 6 \text{ MN m}^{-2}, \quad \text{as before}$$

Example 7.2

The principal stresses at a point in a material are 30 MN m^{-2} and

176

50 MN m^{-2}, both tension. Determine, for a plane inclined at 40° to the plane on which the latter stress acts

(a) the normal and tangential components of stress
(b) the magnitude and direction of the resultant stress.

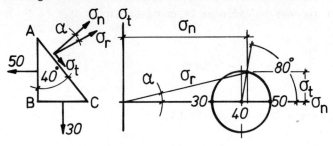

Figure 7.5

The Mohr's circle diagram for the example is shown in figure 7.5. The radius of the circle is 10 MN m^{-2}, thus

$$\sigma_n = 40 + 10 \cos 80° = 41 \cdot 74 \text{ MN m}^{-2}$$

and $\sigma_t = 10 \sin 80° = 9 \cdot 85$ MN m^{-2}

The resultant of these stresses is obtained from the diagram as

$$\sigma_r = \sqrt{(\sigma_n^2 + \sigma_t^2)} = 42 \cdot 9 \text{ MN m}^{-2}$$

The resultant is inclined at an angle α to the direction of σ_n where

$$\alpha = \tan^{-1} \left(\frac{\sigma_t}{\sigma_n}\right) = 13°18'$$

7.3 TORSION COMBINED WITH DIRECT STRESS

Torsion combined with bending or axial load gives rise to a biaxial stress state. As an example of such a combined loading consider a circular-section bar of diameter d subjected to a constant torque T and a constant moment M as shown in figure 7.6.

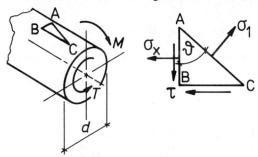

Figure 7.6

177

A small triangular element ABC of the top surface of the bar will experience a tensile stress σ_x and a shear stress τ on AB. BC will be subjected to the complementary shear stress τ alone ($\sigma_y = 0$). It is assumed that AC is a principal plane.

Since the bar is circular in cross-section

$$\tau = \frac{16T}{\pi d^3} \tag{a}$$

and $\sigma_x = \frac{32M}{\pi d^3}$ (b)

where d is the diameter of the bar.

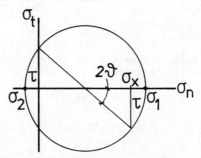

Figure 7.7

Mohr's circle for stress in the element is shown in figure 7.7, thus the maximum principal stress is given by

$$\sigma_1 = \frac{\sigma_x}{2} + \sqrt{\left[\left(\frac{\sigma_x}{2}\right)^2 + \tau^2\right]}$$

or $\sigma_1 = \frac{16}{\pi d^3}[M + \sqrt{(M^2 + T^2)}]$ (c)

The bending moment acting alone which would produce a maximum bending stress in the bar equal to σ_1 is

$$M_{eq} = \frac{1}{2}[M + \sqrt{(M^2 + T^2)}] \tag{d}$$

where M_{eq} is referred to as the equivalent bending moment.

The minimum principal stress in the bar is obtained from figure 7.7 as

$$\sigma_2 = \frac{16}{\pi d^3}[M - \sqrt{(M^2 + T^2)}] \tag{e}$$

thus the maximum shear stress is given by

$$\tau_{max} = \frac{1}{2}(\sigma_1 - \sigma_2) = \frac{16}{\pi d^3}\sqrt{(M^2 + T^2)} \tag{f}$$

178

The torque acting alone which would produce a maximum shear stress in the bar equal to τ_{max} is

$$T_{eq} = \surd(M^2 + T^2) \qquad\qquad\qquad\qquad\qquad\qquad (g)$$

where T_{eq} is referred to as the equivalent torque.

Example 7.3

A hollow circular steel shaft of outside diameter 89 mm and wall thickness 25 mm, transmits 224 kW at 4·17 rev/s and is subject to an axial thrust of 50 kN in addition to a uniform bending moment M. Determine the value of M if the greater principal stress in the shaft is not to exceed 93 MN m^{-2}.

The cross-sectional area A, the second moment of area I and the polar second moment of area J are

$$A = \frac{\pi}{4}(89^2 - 39^2) = 5026 \text{ mm}^2$$

$$I = \frac{\pi}{64}(89^4 - 39^4) = 2966 \times 10^3 \text{ mm}^4$$

and $J = 2I = 5932 \times 10^3$ mm^4

From equation 3.7 we have that the power transmitted by the shaft is

$$P = 2\pi nT = 224 \times 10^3 \text{ watts } (\text{Nm s}^{-1})$$

thus the torque T in the shaft is given by

$$T = \frac{(224)10^3}{2\pi(4\cdot17)} = 8556 \text{ Nm}$$

the surface shear stress is thus

$$\tau = \frac{Tr}{J} = \frac{(8556)(89)10^3}{2(5932)} = 64\cdot2 \text{ MN m}^{-2}$$

The direct stress, which is compressive, is given by

$$\sigma_d = -\frac{50 \times 10^3}{5026} = -9\cdot95 \text{ MN m}^{-2}$$

and for M in MN m, the maximum and minimum bending stresses are

$$\sigma_b = \pm\frac{M(89)10^3}{2(2\cdot966)} = \pm 15\cdot0M \times 10^3 \text{ MN m}^{-2}$$

With the aid of a sketch of the Mohr's circle diagram it can be seen that the greater principal stress in the shaft is compressive and takes the value

$$\sigma_1 = -\left\{\frac{(15M \times 10^3 + 9 \cdot 95)}{2} + \sqrt{\left[\left(\frac{15M \times 10^3 + 9 \cdot 95}{2}\right)^2 + (64 \cdot 2)^2\right]}\right\} \text{ MN m}^{-2}$$

but σ_1 is not to exceed -93 MN m^{-2}, thus

$$M = 2 \cdot 58 \times 10^{-3} \text{ MN m} = 2 \cdot 58 \text{ kN m}$$

7.4 MOHR'S CIRCLE FOR STRAIN

Figure 7.8 shows a small rectangular element of material ABCD which is in a condition of plane strain under direct strains ε_x and ε_y together with a shear strain γ.

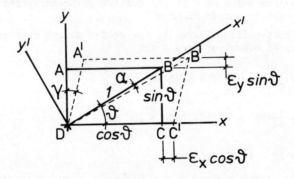

Figure 7.8

We now determine the direct and shear strains on a neighbouring element of the material whose sides (directions x' and y') are initially displaced by an angle θ with respect to the axes x and y. The diagonal DB is of unit length, thus DC $= \cos\theta$ and BC $= \sin\theta$. Under strain, the length of the diagonal is increased by BB'. Since DB is of unit length, BB' represents the direct strain, ε_θ, in the direction of x'. We sum the components of BB' due to the three straining actions, thus

$$\varepsilon_\theta = \varepsilon_x \cos^2\theta + \varepsilon_y \sin^2\theta + \gamma \sin\theta \cos\theta \tag{a}$$

By setting θ equal to $\theta + (\pi/2)$ we obtain the direct strain in the direction of y', thus

$$\varepsilon_{\theta+\pi/2} = \varepsilon_x \sin^2\theta + \varepsilon_y \cos^2\theta - \gamma \sin\theta \cos\theta \tag{b}$$

The total rotation α of the x' axis is obtained by summing the separate rotations due to the three straining actions. Taking positive rotations as clockwise, we have

$$\alpha = \varepsilon_x \sin\theta \cos\theta - \varepsilon_y \sin\theta \cos\theta + \gamma \sin^2\theta \tag{c}$$

By setting θ equal to $\theta + (\pi/2)$ we obtain the rotation β of the y' axis thus

$$\beta = -\varepsilon_x \sin\theta \cos\theta + \varepsilon_y \sin\theta \cos\theta + \gamma \cos^2\theta \tag{d}$$

180

The shear strain, γ_θ, with respect to axes x' and y' is defined as the change of angle between them, thus

$$\gamma_\theta = \beta - \alpha$$

or $\quad \gamma_\theta = -2\varepsilon_x \sin\theta \cos\theta + 2\varepsilon_y \sin\theta \cos\theta + \gamma(\cos^2\theta - \sin^2\theta) \quad\text{(e)}$

Equations a and e may be written in the form

$$\varepsilon_\theta = \frac{1}{2}(\varepsilon_x + \varepsilon_y) + \frac{1}{2}(\varepsilon_x - \varepsilon_y)\cos 2\theta + \frac{1}{2}\gamma \sin 2\theta \qquad (7.3)$$

and $\quad -\dfrac{1}{2}\gamma_\theta = \dfrac{1}{2}(\varepsilon_x - \varepsilon_y)\sin 2\theta - \dfrac{1}{2}\gamma \cos 2\theta \qquad (7.4)$

The close similarity with equations 7.1 and 7.2 will be observed. It follows, therefore, that the Mohr's circle for strain shown in figure 7.9 may be constructed in the same way as the Mohr's circle for stress provided the following substitutions are made

$$\varepsilon_x = \sigma_x, \; \varepsilon_y = \sigma_y, \; \frac{1}{2}\gamma = \tau$$

then $\varepsilon_\theta = \sigma_n$ and $-\dfrac{1}{2}\gamma_\theta = \sigma_t$

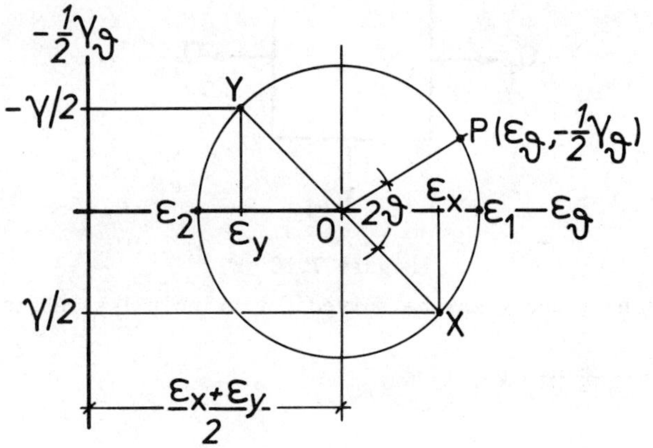

Figure 7.9

A radius vector OP at an angle 2θ measured anticlockwise from OX intersects the circle at P. The coordinates of P are the direct and shear strains referred to axes x' and y'.

The maximum and minimum values of the direct strain are ε_1 and ε_2. These strains are principal strains and occur on principal planes where the shear strain is zero. The maximum shear strain is equal to the difference of the principal strains and occurs on planes which bisect the angle between the principal planes.

The analysis of strain is of vital importance in experimental stress

181

analysis. Strain gauges are used in practice to determine strain changes in loaded structures. A small rosette of three electrical-resistance gauges aligned in different directions may be used to determine strains at a point. It must be noted that the gauges only measure direct strain, but three values of direct strain at three different values of θ enable the Mohr's strain circle to be drawn. From the strain circle it is then possible to determine principal strains and the maximum shear strain.

Principal stresses are related to principal strains. Consider the square element of unit side shown in figure 7.10. The element is subjected to principal stresses σ_1 and σ_2 as shown. The corresponding principal strains are then given by

$$\varepsilon_1 = \frac{\sigma_1}{E} - \nu \frac{\sigma_2}{E} \qquad\qquad\qquad\qquad (f)$$

and $\quad \varepsilon_2 = \frac{\sigma_2}{E} - \nu \frac{\sigma_1}{E} \qquad\qquad\qquad\qquad (g)$

where ν is Poisson's ratio (see chapter 2) and E is Young's modulus.

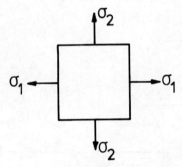

Figure 7.10

Equations f and g may be solved for principal stresses, thus

$$\sigma_1 = \frac{E}{(1 - \nu^2)} (\varepsilon_1 + \nu\varepsilon_2) \qquad\qquad\qquad\qquad (7.5)$$

and $\quad \sigma_2 = \frac{E}{(1 - \nu^2)} (\varepsilon_2 + \nu\varepsilon_1) \qquad\qquad\qquad\qquad (7.6)$

Stresses in any direction at the point may now be determined from the Mohr's stress circle.

Example 7.4

A rosette of three electrical-resistance strain gauges mounted on the surface of a metal plate under stress, gave the following readings

Gauge A, (at 0°) + 0·000592

Gauge B, (at 45°) + 0·000308

Gauge C, (at 90°) - 0·000432

The angles are measured anticlockwise from gauge A. Determine the magnitudes of the principal stresses and their directions relative to gauge A. $E = 200$ GN m^{-2} and $\nu = 0\cdot 30$.

From equation 7.3 it is possible to determine the shear strain associated with the direction of gauge A, thus

$$\varepsilon_B = \tfrac{1}{2}(592 - 432)10^{-6} + \tfrac{1}{2}\gamma_A = 308 \times 10^{-6}$$

thus

$$\gamma_A = 456 \times 10^{-6} \text{ and } \tfrac{1}{2}\gamma_A = 228 \times 10^{-6}$$

This will also be the shear strain associated with the direction of gauge C. It is now possible to construct the strain circle shown in figure 7.11.

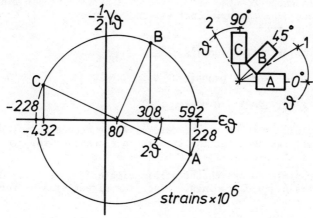

Figure 7.11

The principal strains are thus

$$\varepsilon_1 = [80 + \sqrt{(512^2 + 228^2)}]10^{-6} = 640\cdot 5 \times 10^{-6}$$

and $\varepsilon_2 = [80 - \sqrt{(512^2 + 228^2)}]10^{-6} = -480\cdot 5 \times 10^{-6}$

If the principal plane on which the principal strain ε_1 acts is inclined at an angle θ measured anticlockwise from the direction of gauge A we have

$$\theta = \frac{1}{2}\tan^{-1}\left(\frac{228}{512}\right) = 12°$$

The other principal plane is at $\theta + 90° = 102°$.

From equations 7.5 and 7.6, the principal stresses are

183

$$\sigma_1 = \frac{200 \times 10^3}{1 - (0 \cdot 3)^2}[640 \cdot 5 - (0 \cdot 3)(480 \cdot 5)]10^{-6} \text{ MN m}^{-2}$$

and $\sigma_2 = \frac{200 \times 10^3}{1 - (0 \cdot 3)^2}[- 480 \cdot 5 + (0 \cdot 3)(640 \cdot 5)]10^{-6} \text{ MN m}^{-2}$

thus $\sigma_1 = 190 \cdot 1 \text{ MN m}^{-2}$ (tension)

and $\sigma_2 = -63 \cdot 4 \text{ MN m}^{-2}$ (compression)

The maximum shear stress is given by

$$\tau_{max} = \frac{1}{2}(\sigma_1 - \sigma_2) = 126 \cdot 7 \text{ MN m}^{-2}$$

Example 7.5

A 60° strain gauge rosette is mounted on a steel plate in plane stress. Determine the principal stresses in the plate if the strain gauge readings are

Gauge A (at 0°) + 0·000125

Gauge B (at 60°) + 0·000125

Gauge C (at 120°) + 0·001100

If the plate is 10 mm thick, determine the change in thickness under stress. Assume that the stresses are constant through the thickness of the plate. $E = 200 \text{ GN m}^{-2}$, $\nu = 0 \cdot 3$.

It is possible to sketch the strain circle (figure 7.12) directly from the strain gauge readings. Since the direct strains for gauges

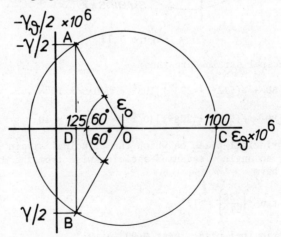

Figure 7.12

A and B are equal, the shear strains associated with these gauges must be equal and opposite. Further, since the angles between the gauges are 60°, the direct strain registered by gauge C must be equal to the maximum principal strain.

From the triangle DOB we have

$$(1100 - \varepsilon_o \times 10^6) \cos 60° = \varepsilon_o \times 10^6 - 125$$

hence

$$\varepsilon_o = 450 \times 10^{-6}$$

and the lesser principal strain is therefore given by

$$\varepsilon_2 = [450 - (1100 - 450)]10^{-6} = -200 \times 10^{-6}$$

Using equations 7.5 and 7.6 to determine principal stresses we have

$$\sigma_1 = + 228 \cdot 6 \text{ MN m}^{-2}$$

and $\sigma_2 = + 28 \cdot 6 \text{ MN m}^{-2}$

The strain in the direction normal to the plane of the plate is given by

$$\varepsilon_2 = -\frac{\nu}{E}(\sigma_1 + \sigma_2) = -386 \times 10^{-6}$$

thus the change in plate thickness is

$$\Delta t = -10 \times 386 \times 10^{-6} = -3 \cdot 86 \times 10^{-3} \text{ mm}$$

Example 7.6

The state of strain at a point on the surface of a high-strength steel structural member under load was measured by means of three electrical-resistance strain gauges. The strain readings and the relative angular positions of the gauges were as follows

Gauge A, (at 0°) + 0·000120

Gauge B, (at 30°) - 0·000063

Gauge C, (at 45°) + 0·000236

Determine the principal stresses and the maximum shear stress at the point of measurement. $E = 200 \text{ GN m}^{-2}$, $\nu = 0 \cdot 3$.

From equation 7.3

$$\varepsilon_B = -63 \times 10^{-6} = 90 \times 10^{-6} + \frac{1}{4}(\varepsilon_y + \sqrt{3}\gamma)$$

and $\varepsilon_C = 236 \times 10^{-6} = 60 \times 10^{-6} + \frac{1}{2}(\varepsilon_y + \gamma)$

185

hence

$$\varepsilon_y = 1669 \times 10^{-6} \text{ and } \gamma = -1317 \times 10^{-6}$$

It is now possible to sketch the strain circle shown in figure 7.13.

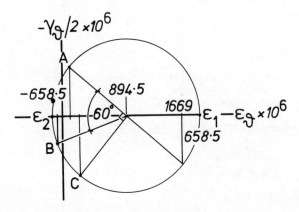

Figure 7.13

From this diagram we have

$$\varepsilon_1 = \{894 \cdot 5 + \sqrt{[(774 \cdot 5)^2 + (658 \cdot 5)^2]}\}10^{-6}$$

$$= 1911 \cdot 1 \times 10^{-6}$$

and $$\varepsilon_2 = \{894 \cdot 5 - \sqrt{[(774 \cdot 5)^2 + (658 \cdot 5)^2]}\}10^{-6}$$

$$= -122 \cdot 1 \times 10^{-6}$$

From equations 7.5 and 7.6, the principal stresses are

$$\sigma_1 = 412 \text{ MN m}^{-2} \text{ and } \sigma_2 = 99 \cdot 2 \text{ MN m}^{-2}$$

Since σ_1 and σ_2 are both positive, the maximum shear stress lies in a plane containing the maximum principal stress and normal to the surface of the member. Its value is (see figure 7.3)

$$\tau_{max} = \frac{1}{2} \sigma_1 = 206 \text{ MN m}^{-2}$$

7.5 A RELATIONSHIP BETWEEN E, G AND ν

In section 2.5.1 we found a relationship between the three elastic constants E, K and ν. We shall now derive a further relationship, this time between E, G and ν.

Consider a thin square element ABCD of unit side subjected to a pure shear stress τ as shown in figure 7.14a. The corresponding pure shear strain is γ. For convenience the distorted element

AB'C'D is shown on the same base (AD) as the undisturbed element. For small strains, the extension of the diagonal is EC'. The direct diagonal strain ε_d, is therefore given by

$$\varepsilon_d = \frac{EC'}{AC} = \left(\frac{\gamma}{\sqrt{2}}\right)\frac{1}{\sqrt{2}} \text{ or } \varepsilon_d = \frac{\tau}{2G} \tag{a}$$

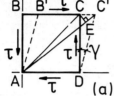 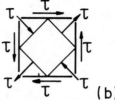

Figure 7.14

We have already seen (section 7.1) that a pure shear system is equivalent to tensile and compressive direct stresses numerically equal to τ acting on planes at 45° to the pure shear planes, (figure 7.14b).

The direct strain of the diagonal AC in terms of the direct stresses is thus

$$\varepsilon_d = \frac{\tau(1 + \nu)}{E} \tag{b}$$

From equations a and b we have

$$E = 2G (1 + \nu) \tag{7.7}$$

If we eliminate ν between equations 2.15 and 7.7, we obtain the following relationship between the three elastic moduli

$$E = \frac{9GK}{G + 3K} \tag{7.8}$$

Example 7.7

A close-coiled helical spring of circular wire and mean diameter 100 mm extends 10·73 mm under an axial load of 12 N. The same spring when firmly fixed at one end was found to rotate through 91·9° when a torque of 6 N m was applied at the other end in a plane at right angles to the spring axis. Use this information to determine Poisson's ratio for the spring material.

When the axial load is applied, the spring wire experiences a constant torque T (= WR). The strain energy stored in the spring is then

$$U = \frac{1}{2} T\theta = \frac{W^2R^2L}{2GJ}$$

but the work done on the spring is given by $W\Delta/2$. Since the work done is equal to the strain energy stored we have

187

$$\frac{1}{2} W\Delta = \frac{W^2 R^2 L}{2GJ}$$

thus

$$\frac{GJ}{L} = \frac{WR^2}{\Delta} = 2796 \text{ N mm} \qquad (1)$$

When the torque is applied, the spring wire experiences a constant moment, M, equal to the torque. Equating strain energy to work done we obtain

$$\frac{1}{2} M\theta = \frac{M^2 L}{2EI}$$

where θ is in radians; thus

$$\frac{EI}{L} = \frac{M}{\theta} = 3741 \text{ N mm} \qquad (2)$$

Since the spring wire is circular, $J = 2I$ and from equations 1 and 2

$$\frac{E}{G} = \frac{3741 \times 2}{2796} = 2 \cdot 676$$

From equation 7.7 we have

$$\frac{E}{G} = 2(1 + \nu)$$

hence $\nu = 0 \cdot 338$.

7.6   STRAIN ENERGY UNDER PLANE STRESS CONDITIONS

Although we are concerned here with biaxial stress states, it is convenient to determine first the strain energy for a triaxial stress system. The result for plane stress is then obtained by setting the normal stress to zero.

7.6.1   Total Strain Energy

Consider the unit cube shown in figure 7.15 subjected to principal stresses σ_1, σ_2 and σ_3 on each of its three mutually perpendicular faces.

The strain ε_1 in the direction of σ_1 is given by

$$\varepsilon_1 = \frac{1}{E}[\sigma_1 - \nu(\sigma_2 + \sigma_3)]$$

the cube is of unit dimensions, thus the extension in the direction of σ_1 is ε_1 and the corresponding force is equal to σ_1.

The component of strain energy stored in the cube due to the principle stress σ_1 is thus

$$U_1 = \frac{\sigma_1}{2E}[\sigma_1 - \nu(\sigma_2 + \sigma_3)]$$

similarly the components of strain energy due to σ_2 and σ_3 are

$$U_2 = \frac{\sigma_2}{2E}[\sigma_2 - \nu(\sigma_1 + \sigma_3)]$$

and $\quad U_3 = \frac{\sigma_3}{2E}[\sigma_3 - \nu(\sigma_1 + \sigma_2)]$

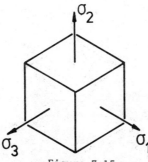

Figure 7.15

The total strain energy stored is the sum of these components, thus

$$U_{total} = \frac{1}{2E}[\sigma_1^2 + \sigma_2^2 + \sigma_3^2 - 2\nu(\sigma_1\sigma_2 + \sigma_1\sigma_3 + \sigma_2\sigma_3)] \qquad (7.9)$$

For plane stress, $\sigma_3 = 0$, hence

$$U'_{total} = \frac{1}{2E}[\sigma_1^2 + \sigma_2^2 - 2\nu\sigma_1\sigma_2] \qquad (7.10)$$

7.6.2 Distortional, or Shear Strain Energy

The triaxial system of principal stresses acting on the cube in figure 7.15, may be split into the two systems shown in figure 7.16.

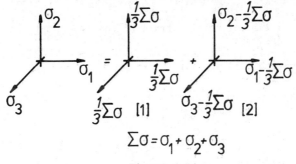

$$\Sigma\sigma = \sigma_1 + \sigma_2 + \sigma_3$$

Figure 7.16

The uniform stress or hydrostatic pressure of system 1 produces

189

a volume change but no distortion, (the cube is strained into a cube). The stresses of system 2 produce distortion only with no change in the volume of the cube. That this is so can be proved if we consider a unit cube subjected to stress system 2.

The principal strains are

$$\varepsilon_1 = \frac{1}{E}[\sigma_1 - \frac{1}{3}\Sigma\sigma] - \frac{\nu}{E}[\sigma_2 + \sigma_3 - \frac{2}{3}\Sigma\sigma]$$

$$\varepsilon_2 = \frac{1}{E}[\sigma_2 - \frac{1}{3}\Sigma\sigma] - \frac{\nu}{E}[\sigma_1 + \sigma_3 - \frac{2}{3}\Sigma\sigma] \qquad (a)$$

and $$\varepsilon_3 = \frac{1}{E}[\sigma_3 - \frac{1}{3}\Sigma\sigma] - \frac{\nu}{E}[\sigma_1 + \sigma_2 - \frac{2}{3}\Sigma\sigma]$$

where

$$\Sigma\sigma = \sigma_1 + \sigma_2 + \sigma_3$$

The volume, V_1, of the strained cube is given by

$$V_1 = 1 + \Delta V = (1 + \varepsilon_1)(1 + \varepsilon_2)(1 + \varepsilon_3)$$

$$= 1 + \varepsilon_1 + \varepsilon_2 + \varepsilon_3$$

since products of small strains may be ignored, hence

$$\Delta V = \varepsilon_1 + \varepsilon_2 + \varepsilon_3 = \Sigma\varepsilon$$

but from equations a above $\Sigma\varepsilon = 0$, thus

$$\Delta V = 0$$

The strain energy stored in the cube by the stresses of system 1 is given by

$$U_h = \frac{1}{2}\left(\frac{\Sigma\sigma}{3}\right)\frac{dV}{V}$$

where dV/V is the volumetric strain of the cube, and $\Sigma\sigma/3$ is the hydrostatic pressure.

From equation 2.14 we have that

$$\frac{dV}{V} = \frac{1}{3K}\Sigma\sigma$$

thus $$U_h = \frac{1}{18K}(\Sigma\sigma)^2$$

but from equation 2.15

$$E = 3K(1-2\nu)$$

190

thus $U_h = \dfrac{(1 - 2\nu)}{6E}(\sigma_1 + \sigma_2 + \sigma_3)^2$

The total strain energy stored in the cube by the system of principal stresses σ_1, σ_2 and σ_3 has already been obtained in equation 7.9, thus the strain energy of distortion or the shear strain energy is given by

$$U_s = U_{total} - U_h$$

thus $U_s = \dfrac{(1 + \nu)}{3E}[\sigma_1^2 + \sigma_2^2 + \sigma_3^2 - (\sigma_1\sigma_2 + \sigma_1\sigma_3 + \sigma_2\sigma_3)]$ \qquad (7.11)

For plane stress, $\sigma_3 = 0$ hence

$$U_s' = \dfrac{(1 + \nu)}{3E}[\sigma_1^2 + \sigma_2^2 - \sigma_1\sigma_2] \qquad (7.12)$$

7.7 THEORIES OF ELASTIC FAILURE

For a uniaxially stressed system, elastic behaviour may be assumed provided the stress is less than the limit for which stress is proportional to strain. A commonly used term for the elastic limit is the yield stress, σ_y, although yielding is strictly a phenomenon associated with annealed mild steel. It is assumed that for ductile materials, σ_y is the same in simple tension and compression. For brittle materials such as cast iron and concrete, the tensile yield stress is much smaller than the compressive yield stress.

In a biaxial stress system the separate contributions of the two principal stresses to elastic failure are less easy to evaluate. In the past a number of theories have been suggested to explain elastic failure under biaxial stress, the most important of which have withstood the test of comparison with experimental results. We will now examine four of these theories of elastic failure.

7.7.1 The Maximum Principal Stress Theory

This theory assumes that elastic failure will occur when the maximum principal stress is equal to the yield stress in simple tension or compression, thus

$$\sigma_1 = \sigma_y, \text{ or } \sigma_2 = \sigma_y \qquad (a)$$

The theory gives good predictions for brittle materials (which are weak in tension) if the maximum principal tensile stress is limited to the uniaxial tensile yield stress.

7.7.2 The Maximum Shear Stress Theory

It is assumed that elastic failure occurs when the maximum shear stress in the biaxial stress system is equal to the shear stress,

$\sigma_y/2$, which corresponds to failure in simple tension. Thus for σ_1 and σ_2 having the same sign

$$\sigma_1 = \sigma_y \text{ or } \sigma_2 = \sigma_y \qquad \text{(b)}$$

If σ_1 and σ_2 are of opposite sign

$$\sigma_1 - \sigma_2 = \sigma_y \text{ or } \sigma_2 - \sigma_1 = \sigma_y \qquad \text{(c)}$$

The maximum shear stress theory, sometimes referred to as the Tresca theory, gives good results for ductile materials. It is often used because of its simplicity.

7.7.3 The Maximum Total Strain Energy Theory

This theory (due to Haigh) assumes that elastic failure occurs when the maximum total strain energy in a biaxial stress system is equal to the total strain energy corresponding to failure in simple tension. From equations 7.10 and 6.5 we have therefore

$$\frac{1}{2E}[\sigma_1^2 + \sigma_2^2 - 2\nu\sigma_1\sigma_2] = \frac{\sigma_y^2}{2E}$$

or $\quad \sigma_1^2 + \sigma_2^2 - 2\nu\sigma_1\sigma_2 = \sigma_y^2 \qquad \text{(d)}$

The theory is not often used although it gives reasonably good results for ductile materials.

7.7.4 The Maximum Shear Strain Energy Theory

It is assumed that elastic failure occurs in a biaxial stress system when the maximum shear strain energy is equal to the shear strain energy at failure in simple tension. From equation 7.12 we have

$$U_s' = \frac{(1 + \nu)}{3E}[\sigma_1^2 + \sigma_2^2 - \sigma_1\sigma_2]$$

The uniaxial shear strain energy is obtained from equation 7.12 by setting $\sigma_1 = \sigma_y$ and $\sigma_2 = 0$. Thus for elastic failure

$$\sigma_1^2 + \sigma_2^2 - \sigma_1\sigma_2 = \sigma_y^2 \qquad \text{(e)}$$

This theory is widely used for ductile materials. It provides good agreement with experimental results.

7.7.5 Graphical Representation of the Theories of Failure

Figure 7.17 shows a diagram on which the limiting principal stresses predicted by the four failure theories (equations a to e) are plotted with reference to axes σ_1 and σ_2.

Elastic behaviour may be assumed for combinations of σ_1 and σ_2 within the boundary imposed by a particular theory. Elastic failure occurs at the boundary.

192

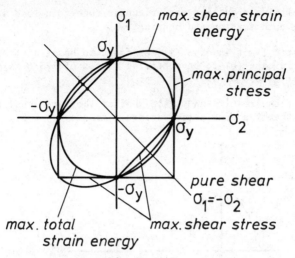

Figure 7.17

7.7.6 Yielding in Pure Shear

If a material has reached yield in pure shear, we have

$$\sigma_1 = -\sigma_2 = \tau_y \qquad\qquad\qquad (f)$$

where τ_y is the shear yield stress.

If we substitute for σ_1 and σ_2 in equations a, c, d and e it is possible to determine the shear yield stress in terms of the yield stress in simple tension, thus if $\nu = 0 \cdot 3$

(i) maximum principal stress

$$\tau_y = \sigma_y$$

(ii) maximum shear stress

$$\tau_y = 0 \cdot 5 \ \sigma_y$$

(iii) maximum total strain energy

$$\tau_y = 0 \cdot 62 \ \sigma_y$$

(iv) maximum shear strain energy

$$\tau_y = 0 \cdot 58 \ \sigma_y$$

Example 7.8

A sample of steel was tested (a) by direct tension of a solid bar and

(b) by submitting a cantilevered circular tube to a load at the free end causing both torsion and bending.

The elastic limit in case (a) was found to be at a stress of 263 MN m^{-2} and in case (b) to be a bending stress of 123·6 MN m^{-2} together with a shear stress of 117·5 MN m^{-2}.

Examine these results and state whether they are consistent with any of the theories of elastic failure. Assume $\nu = 0·3$.

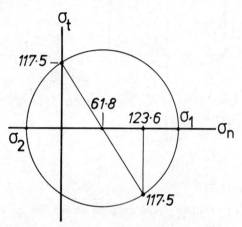

Figure 7.18

Figure 7.18 shows the Mohr's stress circle for case b, thus

$$\sigma_1 = 61·8 + \sqrt{[(61·8)^2 + (117·5)^2]} = 194·6 \text{ MN m}^{-2}$$

and $\quad \sigma_2 = 61·8 - \sqrt{[(61·8)^2 + (117·5)^2]} = -71·0 \text{ MN m}^{-2}$

(i) From the maximum principal stress theory
$\sigma_y = \sigma_1 = 194·6 \text{ MN m}^{-2}$

(ii) From the maximum shear stress theory
$\sigma_y = \sigma_1 - \sigma_2 = 265·6 \text{ MN m}^{-2}$

(iii) From the maximum total strain energy theory
$\sigma_y = \sqrt{(\sigma_1^2 + \sigma_2^2 - 2\nu\sigma_1\sigma_2)} = 226·3 \text{ MN m}^{-2}$

(iv) From the maximum shear strain energy theory
$\sigma_y = \sqrt{(\sigma_1^2 + \sigma_2^2 - \sigma_1\sigma_2)} = 238·2 \text{ MN m}^{-2}$

Clearly, the maximum shear stress theory provides the closest approximation to the true yield stress for the material.

Example 7.9

A tube of mean diameter 75 mm and wall thickness 2·5 mm is made of mild steel with an elastic limit of 230 MN m^{-2} in simple tension. Calculate the torque that may be transmitted by the tube if the safety factor against yield is 2·5 and the failure criterion is given by

(a) the maximum shear stress theory

(b) the maximum total strain energy theory, and

(c) the maximum shear strain energy theory

Take $\nu = 0·3$.

The tube is thin, thus under a torque of T Nm, the shear stress is given by

$$\tau = \frac{T}{2\pi R^2 t} = \frac{2T \times 10^3}{\pi (75)^2 (2·5)} \text{ MN m}^{-2}$$

or $\tau = 0·0453\ T \text{ MN m}^{-2}$

Since the tube is in pure shear, the principal stresses are

$$\sigma_1 = -\sigma_2 = 0·0453\ T \text{ MN m}^{-2}$$

The limiting value of stress in direct tension is given by

$$\sigma_{lim} = \frac{230}{2·5} = 92 \text{ MN m}^{-2}$$

Substituting σ_{lim} for yield stress we obtain (from section 7.7.6) for the three theories of failure

(a) $0·0453\ T = 0·5(92)$ or $T = 1015$ Nm

(b) $0·0453\ T = 0·62(92)$ or $T = 1259$ Nm

(c) $0·0453\ T = 0·58(92)$ or $T = 1178$ Nm

Example 7.10

A solid circular shaft 100 mm diameter is subjected to combined bending and torsion. The magnitude of the bending moment is three times that of the torque. If the direct tension yield point of the material is 370 MN m^{-2} and the factor of safety on yield is to be 4, calculate the allowable torque by

(a) the maximum principal stress theory

(b) the maximum shear stress theory

(c) the maximum shear strain energy theory

If the torque in the shaft is T kN m, the surface shear stress, τ and the maximum bending stress, σ are given by

$$\tau = \frac{16T}{\pi} \text{ MN m}^{-2}$$

and $\sigma = \dfrac{96T}{\pi}$ MN m^{-2}

The principle stresses are thus

$$\sigma_1 = \frac{48T}{\pi}\left(1 + \frac{\sqrt{10}}{3}\right) = 31 \cdot 4 \; T \text{ MN m}^{-2}$$

and $\sigma_2 = \dfrac{48T}{\pi}\left(1 - \dfrac{\sqrt{10}}{3}\right) = -0 \cdot 83 \; T \text{ MN m}^{-2}$

The limiting value of stress in direct tension is given by

$$\sigma_{\text{lim}} = \frac{370}{4} = 92 \cdot 5 \text{ MN m}^{-2}$$

The maximum allowable torques from the three failure theories are thus

(a) $31 \cdot 4 \; T = 92 \cdot 5$ MN m^{-2} or $T = 2 \cdot 94$ kN m

(b) $31 \cdot 4 \; T + 0 \cdot 83 \; T = 92 \cdot 5$ MN m^{-2} or $T = 2 \cdot 87$ kN m

(c) $(31 \cdot 4 \; T)^2 + (0 \cdot 83 \; T)^2 - (31 \cdot 4 \; T)(-0 \cdot 83 \; T) = (92 \cdot 5)^2$ or $T = 2 \cdot 91$ kN m

7.8 PROBLEMS FOR SOLUTION

1. At a point in a member there are tensile stresses of 50 and 25 MN m^{-2} at right angles together with shear stresses of 20 MN m^{-2}. The directions of the shear stresses are shown in figure 7.1. Find

(a) normal and tangential stresses on the plane at 30° anti-clockwise to the plane of the 50 MN m^{-2} stress

(b) principal stresses and planes

(c) planes and value of maximum shear stress.

Consider stresses in two dimensions only.
(61·07 and 0·82 MN m^{-2}; 61·08 and 13·9 MN m^{-2}; 29° and 119°; 74° and 164°, 23·6 MN m^{-2})

2. A right-angled triangle ABC with the right angle at C represents planes in an elastic material. There are shear stresses of 30 MN m^{-2} acting along planes AC and BC towards C combined with normal tensile stresses of 50 and 40 MN m^{-2} respectively. Considering stresses in the plane ABC only, determine the angle of plane AB measured anti-clockwise relative to plane AC when the resultant stress on AB has

(a) the greatest magnitude
(b) the least magnitude
(c) the greatest component along AB
(d) the least inclination to AB

(40°15', 130°15', 85°15' and 175°15', 106°30')

196

3. A solid circular-section cantilever of diameter 100 mm carries a thrust of 5 kN in the manner shown in figure 7.19 together with a torque of 0·5 kN m. Determine the greatest values of tensile and compressive stress in the cantilever and show on a sketch where they occur.

(3·7 MN m^{-2}, bottom; -4·6 MN m^{-2}, top)

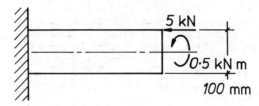

Figure 7.19

4. A solid circular-section shaft of diameter D carries at its free end a moment M and a torque αM. If the loading is such that the maximum principal stress in the shaft does not exceed half the material yield stress in tension, σ_y, determine M in terms of D, α and σ_y.

$$\left(\frac{\pi D^3 \sigma_y}{32}[1 + \sqrt{(1 + \alpha^2)}]^{-1} \right)$$

5. At a point in a material subjected to plane stress, the stresses on a certain plane are 75 MN m^{-2} tension and 50 MN m^{-2} shear. On another plane stresses are 45 MN m^{-2} tension and 40 MN m^{-2} shear as shown in figure 7.20. Find the angle between the planes, the principal stresses and the angles between the principal planes and the plane carrying the 75 MN m^{-2} tension.

(108·4°, 125 and 25 MN m^{-2}, 45° and 135°)

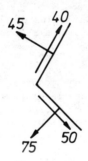

Figure 7.20

6. An element of elastic material is acted upon by three principal stresses and the three principal strains are measured. Show that the principal stress in direction x is given by

$$\sigma_x = \frac{E}{(1 + \nu)}\left(\epsilon_x + \frac{\nu}{1 - 2\nu} \Delta \right)$$

197

where Δ, the volumetric strain, is the sum of the principal strains, and ν is Poisson's ratio.

In a certain test, the principal strains were found to be $+ 710 \times 10^{-6}$, $+ 1400 \times 10^{-6}$ and $- 1850 \times 10^{-6}$. Determine the three principal stresses. Take $E = 200$ GN m^{-2}, $\nu = 0\cdot3$.
$(+139\cdot2, +245\cdot3, -254\cdot5$ MN m$^{-2})$

7. A plaster cube made in a steel mould with an open top was observed to behave as follows; during setting the cube showed a steadily increasing vertical strain which finally ceased at a value of 64×10^{-6}. Immediately the cube was removed from the mould, this strain dropped to a value ε. If Poisson's ratio for the plaster is $0\cdot28$, what value of ε would confirm that the strain of the mould was negligible?
(36×10^{-6})

8. A hollow steel shaft 300 mm outside diameter and 150 mm inside diameter has two strain gauges attached to it at right angles to each other and at $45°$ to the axis of the shaft. At a shaft speed of 3 rev/s, one gauge indicated a strain of $+ 500 \times 10^{-6}$ and the other a strain of $- 500 \times 10^{-6}$. Determine the power being transmitted by the shaft. $E = 200$ GN m^{-2}.
$(9\cdot37$ MW$)$

9. A rectangular steel plate 12 mm thick is subjected to mutually perpendicular stresses σ_x and σ_y acting normally to the edges of the plate. Strain gauges record that the tensile strains in the x- and y-directions are $+300 \times 10^{-6}$ and $+ 70 \times 10^{-6}$ respectively. Determine the change in plate thickness and the stresses on a plane making an angle of $30°$ with the y-axis. $E = 200$ GN m^{-2} and $\nu = 0\cdot3$.
$(-1\cdot9 \times 10^{-3}$ mm, $61\cdot7$ and $15\cdot4$ MN m$^{-2})$

10. A strain gauge rosette has three gauges set at $0°$, $45°$ and $90°$ to the x-direction at a point on a steel component in plane stress. The gauges record strains of $+ 350 \times 10^{-6}$, $+ 700 \times 10^{-6}$ and $- 70 \times 10^{-6}$ respectively. Determine the angles of the principal planes relative to the x- direction and the maximum shear stress at the point. $E = 200$ GN m^{-2} and $\nu = 0\cdot3$.
$(33°24'$ and $123°24'$, $93\cdot8$ MN m$^{-2})$

11. An equiangular strain gauge rosette has gauges at $0°$, $60°$ and $120°$ to the x-direction. The gauge readings were $+ 400 \times 10^{-6}$, $+ 297 \times 10^{-6}$ and $- 397 \times 10^{-6}$. Assuming plane stress, determine the principal stresses and the angles of the principal planes relative to the x-direction.
$E = 200$ GN m^{-2} and $\nu = 0\cdot3$.
$(105\cdot5$ MN m^{-2}, $-48\cdot3$ MN m^{-2}, $26°34'$, $116°34')$

12. Two cylindrical test specimens, 15 mm in diameter are made from a sample of material. A tensile test on one gave an elongation of $41\cdot7\times10^{-3}$ mm on a gauge length of $50\cdot8$ mm under an axial load of 10 kN. A torsion test on the other gave an angular deflexion of $2\cdot27 \times 10^{-3}$ radians over the same gauge length when the torque was 6 Nm. Use

these results to determine the four elastic constants E, G, K and ν for the material.
(69·0, 27·0 and 52·3 GN m^{-2}, 0·28)

13. A bolt under an axial tensile load of 10 kN is submitted to a shear force of 5 kN. Assuming that the shear force is distributed uniformly over the cross-section of the bolt, determine a safe diameter if the appropriate theories of elastic failure for the material are those due to Tresca and von Mises (maximum shear stress and maximum shear strain energy theories respectively). The elastic limit in direct tension is 250 MN m^{-2} and the safety factor on yield is to be 4.
(16·4 mm)

14. A solid circular shaft, 50 mm diameter is subjected to a bending moment of 2 kN m, an axial thrust of 50 kN and a torque T. The material of the shaft has a compressive yield stress of 280 MN m^{-2}. Determine the limiting torque to cause elastic failure according to

(a) the maximum shear stress theory

(b) the maximum strain energy theory

(c) the maximum shear strain energy theory

Poisson's ratio for the material is 0·34.
(2·55, 3·12 and 2·95 kN m)

8 SHEAR EFFECTS IN BEAMS

Where previously we have had to consider shear stresses produced by shear forces, we have assumed for simplicity that they were uniformly distributed over the section depth. This is not the case and in this chapter we investigate a more exact picture of the actual distribution.

We shall be discussing an approximate theory of shear stress distribution which assumes that the simple theory of bending (chapter 4) applies when shear forces act.

8.1 THE DISTRIBUTION OF SHEAR STRESS IN BEAMS

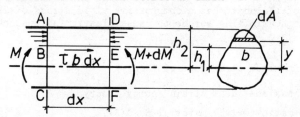

Figure 8.1

Consider an element ABDE (figure 8.1) in a uniform beam having a vertical axis of symmetry. The element is obtained by making two transverse cuts ABC and DEF , dx apart, and a horizontal cut BE parallel to and distance h_1 above the neutral axis. The beam is subjected to a varying bending moment so that on face ABC we have a moment M and on face DEF, the moment is $M + dM$.

We assume that the simple theory of bending applies here although it is only strictly true for pure bending (where plane sections remain plane) thus the longitudinal force on a small area dA of face AB is given by

$$dP = \frac{My}{I}\, dA$$

where I is the second moment of area of the whole cross-section about the neutral axis.

The total longitudinal force on face AB is thus

$$P_{AB} = \frac{M}{I} \int_{h_1}^{h_2} y\ dA \qquad\qquad (a)$$

Similarly the total longitudinal force on face DE is

$$P_{DE} = \frac{M + dM}{I} \int_{h_1}^{h_2} y\ dA \qquad\qquad (b)$$

The two forces P_{AB} and P_{DE} differ in magnitude. The equilibrium of the element ABDE can only be assured by a balancing force due to a shear stress on the horizontal face BE. For reasons that we will discuss later, no longitudinal shear stress can exist on the free surface AD.

If τ is the longitudinal shear stress on BE which is assumed constant across the width, we have for equilibrium of the element.

$$\tau b dx = P_{DE} - P_{AB}$$

or $\quad \tau = \dfrac{dM}{dx} \times \dfrac{1}{Ib} \displaystyle\int_{h_1}^{h_2} y \, dA$ $\qquad\qquad$ (c)

but from chapter 1

$$\frac{dM}{dx} = Q$$

where Q is the shear force at the element.

The integral in equation c we denote by $A\bar{y}$, where A is the area of the cross-section above BE and $\bar{y}$ is the distance of its centroid from the neutral axis. Thus the longitudinal shear stress becomes

$$\tau = \frac{QA\bar{y}}{Ib} \qquad\qquad (8.1)$$

The longitudinal shear force per unit length of beam, or the shear flow (see section 6.3) is given by

$$q = \tau b = \frac{QA\bar{y}}{I} \qquad\qquad (8.2)$$

The longitudinal shear stress τ is accompanied by a complementary vertical shear stress of equal magnitude, thus equation 8.1 gives the value of the vertical shear stress at a height h_1 above the neutral axis.

Since there is no shear stress on a free surface, the vertical shear stress cannot have a component that is normal to the boundary of the cross-section. The vertical and longitudinal shear stresses at the top and bottom of the beam must therefore be zero.

Example 8.1

Determine the distribution of vertical shear stress due to a vertical shear force Q for

(i) a rectangular section

(ii) a square section with a vertical diagonal

(i) The rectangular section. From figure 8.2a, the area above AB is given by

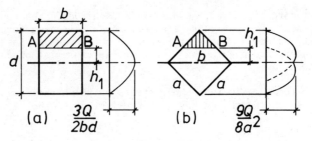

Figure 8.2

$$A = b\left(\frac{d}{2} - h_1\right) = \frac{bd}{2}\left(1 - \frac{2h_1}{d}\right)$$

The distance of the centroid of this area from the neutral axis is

$$\bar{y} = \frac{d}{4}\left(1 + \frac{2h_1}{d}\right)$$

The second moment of area is

$$I = \frac{bd^3}{12}$$

thus from equation 8.1

$$\tau = \frac{3Q}{2bd}\left[1 - \left(\frac{2h_1}{d}\right)^2\right] \tag{1}$$

The distribution of shear stress is therefore parabolic in form as shown in figure 8.2a. The maximum shear stress is 1·5 times the mean shear stress and it occurs at the neutral axis.

(ii) The square section with a vertical diagonal. From figure 8.2b, the breadth of the section at AB is given by

$$b = a\sqrt{2}\left(1 - \sqrt{2}\frac{h_1}{a}\right)$$

thus the area above AB is

$$A = \frac{a^2}{2}\left(1 - \frac{h_1\sqrt{2}}{a}\right)^2$$

and the distance of the centroid of this area from the neutral axis is given by

$$\bar{y} = \frac{a}{3\sqrt{2}}\left(1 + \frac{2h_1\sqrt{2}}{a}\right)$$

The second moment of area for a square section about any axis through the centroid is given by

202

$$I = \frac{a^4}{12}$$

thus from equation 8.1

$$\tau = \frac{Q}{a^2}\left(1 - \frac{\sqrt{2}h_1}{a}\right) \times \left(1 + \frac{2\sqrt{2}h_1}{a}\right) \tag{2}$$

The shear stress is zero for h_1 equal to $a/\sqrt{2}$ which is as expected. Equation 2 also gives zero shear stress for h_1 equal to $-a/2\sqrt{2}$. This result is due to the discontinuity in the shape of the cross-section at the neutral axis. Equation 2 therefore holds only for positive values of h_1. If we repeat the calculation for the lower half of the section and superimpose the two stress distributions, we obtain the final distribution shown in figure 8.2b. The maximum stress in the section occurs when h_1 is equal to $\pm a/4\sqrt{2}$ and takes the value $9Q/8a^2$, or 1·125 times the mean shear stress.

The square with a vertical diagonal is one of the few sections in which the maximum shear stress does not occur at the neutral axis.

Example 8.2

Determine the distribution of vertical shear stress due to a vertical shear force Q at a point in a beam of solid circular cross-section having a radius R. Hence show that the maximum shear stress in a thin circular tube is twice the mean shear stress.

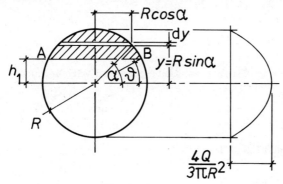

Figure 8.3

From figure 8.3 we have

$$y = R \sin \alpha$$

thus $dy = R \cos \alpha \, d\alpha$

and $dA = 2R^2 \cos^2 \alpha \, d\alpha$

therefore

$$A\bar{y} = \int_{\theta}^{\pi/2} 2R^3 \sin \alpha \cos^2 \alpha \, d\alpha$$

203

or $\quad A\bar{y} = -2R^3 \int_\theta^{\pi/2} \cos^2 \alpha \ d(\cos \alpha) = \frac{2R^3}{3} \cos^3 \theta$ $\qquad$ (1)

From equation 8.1 we have

$$\tau = \frac{4Q}{3\pi R^2} \cos^2 \theta \qquad\qquad (2)$$

When θ is zero, we have the maximum shear stress in the section, thus

$$\tau_{max} = \frac{4Q}{3\pi R^2} = \frac{4}{3} \tau_{mean}$$

Equation 2 gives the magnitude of the vertical components of the shear stress across AB on the assumption that they are uniformly distributed. At the intersection with the vertical centreline, this is also the total shear stress. At A and B on the surface of the beam, the total shear stress is tangential to the boundary thus horizontal shear stresses are distributed along AB, having a maximum value at the surface of the beam and falling to zero at the centre. When AB coincides with the horizontal centreline, the horizontal components of shear stress are all zero.

At the horizontal centreline of a thin tube of outside diameter R and wall thickness t, we have from equation 1 above that

$$A\bar{y} = \frac{2}{3} [R^3 - (R - t)^3]$$

also $I = \frac{\pi}{4}[R^4 - (R - t)^4]$

Since the tube is thin, t is small compared with R thus

$$A\bar{y} \simeq 2R^2 t$$

and $I \simeq \pi R^3 t$

hence from equation 8.1

$$\tau_{max} = \frac{2QR^2 t}{2\pi R^3 t^2} = \frac{Q}{\pi R t}$$

but $\quad \tau_{mean} = \frac{Q}{2\pi R t}$

thus $\tau_{max} = 2\tau_{mean}$

Example 8.3

Two 229 mm × 76 mm channels are placed back to back and the top flanges are connected by a plate 225 mm wide and 16 mm thick. This compound section forms a beam 1·5 m long and carries a uniformly distributed load of intensity 260 kN m^{-1}. If each flange is connected

204

to the plate by two equal fillet welds, determine the maximum shear force per unit length that the welds are required to resist. The major second moment of area and the cross-sectional area for each channel are respectively, 2600 cm^4 and 33 cm^2.

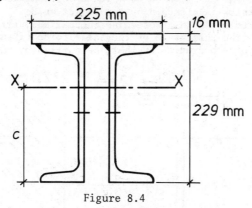

Figure 8.4

Figure 8.4 shows the cross-section of the compound beam. If the position of the neutral axis is at a distance c from the outside of the bottom flanges of the channels, we have

$$c = \frac{2(114 \cdot 5)(3300) + (225)(16)(237)}{2(3300) + (225)(16)} \text{ mm}$$

or $c = 157 \cdot 7$ mm

The second moment of area for the section about the neutral axis is given by

$$I_x = 2(2600 \times 10^4) + 2(3300)(157 \cdot 7 - 114 \cdot 5)^2$$

$$+ \frac{(225)(16)^3}{12} + (225)(16)(237 - 157 \cdot 7)^2 \text{ mm}^4$$

or $I_x = 87 \cdot 0 \times 10^6$ mm^4

The maximum shear force occurs at the ends of the beam thus

$$Q = \frac{1 \cdot 5 \times 260}{2} = 195 \text{ kN}$$

The longitudinal shear force per unit length of beam (or shear flow) at the welded connexion is given by equation 8.2, thus

$$q = \frac{(195 \times 10^3)(225)(16)(237 - 157 \cdot 7)}{(87 \cdot 0 \times 10^6)}$$

or $q = 641 \cdot 5$ kN m^{-1}

There are four welds, thus each weld is required to carry a longitudinal shear force of $160 \cdot 4$ kN m^{-1}.

205

Example 8.4

A tee-section beam symmetrical about a vertical axis has a top flange 100 mm wide and 12 mm thick to which a vertical web plate 160 mm deep and 10 mm thick is attached by two fillet welds. The section carries a shear force of 40 kN. Determine the maximum vertical shear stress in the web, the percentage of the shear force taken by the web and the shear force per unit length to be carried by each weld.

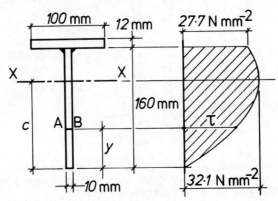

Figure 8.5

Figure 8.5 shows the cross-section of the tee beam together with the distribution of vertical shear stress in the web. If the net neutral axis is a distance c above the bottom of the web plate, we have

$$c = \frac{(12)(100)(166) + (160)(10)(80)}{(12)(100) + (160)(10)} \text{ mm}$$

or $c = 116 \cdot 8$ mm

The second moment of area about the neutral axis is therefore given by

$$I_x = \frac{(100)(12)^3}{12} + (100)(12)(166 - 116 \cdot 8)^2$$

$$+ \frac{(10)(160)^3}{12} + (10)(160)(116 \cdot 8 - 80)^2 \text{ mm}^4$$

or $I_x = 8 \cdot 5 \times 10^6$ mm^4

At a horizontal section through the web at a distance y above the bottom of the web, we have

$$A\bar{y} = (10)y\left(116 \cdot 8 - \frac{y}{2}\right) = 5y(233 \cdot 7 - y) \text{ mm}^3$$

Thus from equation 8.1 the vertical shear stress in the web is given by

206

$$\tau = \frac{(40 \times 10^3)}{(8 \cdot 5 \times 10^6)(10)} \; 5y \; (233 \cdot 7 - y) \; \text{N mm}^{-2}$$

or $\quad \tau = 2 \cdot 35y \; (233 \cdot 7 - y) 10^{-3} \; \text{N mm}^{-2}$

This equation enables the stress distribution in figure 8.5 to be drawn. The maximum vertical shear stress occurs at the neutral axis, thus

$$\tau_{max} = 32 \cdot 1 \; \text{N mm}^{-2}$$

The vertical shear stress at the web to flange junction ($y = 160$ mm) is given by

$$\tau_1 = 27 \cdot 7 \; \text{N mm}^{-2}$$

and the shear flow at this point is

$$q = \tau_1(10) = 277 \; \text{N mm}^{-1}$$

The shear force per unit length to be resisted by each weld is therefore 138 kN m^{-1}.

The total shear force in the web is given by

$$Q_w = \int_0^{160} 10\tau \; dy \; \text{N}$$

hence

$$Q_w = 38 \cdot 2 \; \text{kN}$$

Thus 95·5% of the shear force is carried by the web. Since the web thickness is small, the vertical shear stress may reasonably be assumed to have a uniform horizontal distribution. This is not the case for the flange. The determination of the distribution of vertical shear stress in wide plates is outside the scope of this book; but for the problems we shall be discussing the effect is not important.

8.2 SHEAR FLOW IN THIN-WALLED OPEN SECTIONS

Many common structural sections may be treated as thin-walled open sections. In this category we include I- and channel sections for example but not rectangular or circular tubes. For thin-walled sections we assume that the direction of the shear stress is parallel to the section boundary and ignore the distribution of shear stress in the direction normal to the boundary.

Figure 8.6 shows an element of a thin-walled beam subjected to moments M and $M + dM$ on the two end faces. By an identical argument to that which led to the derivation of equation 8.1 we see that the longitudinal shear stress τ at the section AB is given by

$$\tau = \frac{dM}{dx} \times \frac{A\overline{y}}{It} = \frac{QA\overline{y}}{It}$$

The shear flow parallel to the boundary is therefore given by equation 8.2, thus

$$q = \frac{QA\overline{y}}{I}$$

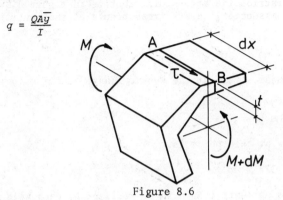

Figure 8.6

Example 8.5

Determine the shear flow in an *I*-section subjected to a shear force *Q*. The depth of the section is *D* and the flange breadth is *B*. The thicknesses of the web and flange are respectively *t* and *T*. Treat the section as thin.

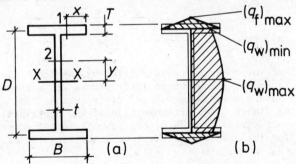

Figure 8.7

Figure 8.7a shows the cross-section. In the flanges, $\overline{y}$ is constant and takes the value $D/2$ thus to the right of section 1 in the flanges we have

$$\overline{Ay} = \frac{xDT}{2}$$

thus the shear flow in the flanges is obtained from equation 8.2 as

$$q_f = \frac{QDT}{2I_x} x \qquad\qquad (1)$$

At section 2 (in the web) we have

208

$$A\bar{y} = \frac{D}{2} TB + \frac{t}{2}\left[\left[\frac{D}{2}\right]^2 - y^2\right]$$

thus the shear flow in the web is

$$q_w = \frac{Q}{2I_w}\left\{DTB + \tau\left[\left[\frac{D}{2}\right]^2 - y^2\right]\right\} \tag{2}$$

The shear flow distributions shown in figure 8.7b are drawn from equations 1 and 2. In the flanges, the shear flow varies linearly from zero at the toes to a maximum at the web to flange junction given by

$$(q_f)_{max} = \frac{QDTB}{4I_x}$$

In the web, the shear flow distribution is parabolic. At the web to flange junction we have the minimum value given by

$$(q_w)_{min} = \frac{QDTB}{2I_x}$$

The maximum shear flow in the web is at the neutral axis and takes the value

$$(q_w)_{max} = \frac{QD}{8I_x}(4BT + Dt)$$

The shear stress at any point in the section is obtained by dividing the shear flow by the appropriate thickness.

From equation 2 it can be checked that as a consequence of assuming the section to be thin, all the vertical shear force is taken by the web.

Example 8.6

The top-hat section shown in figure 8.8 a is made by bending up thin plate of thickness t. Determine the distribution of shear flow under a downward vertical shear force Q.

By inspection, the centroid of the section is seen to be at mid-height. The shear flow is proportional to $A\bar{y}$, thus starting with section 1 in the lower flange

$$(A\bar{y}) = tx_1\left(-\frac{b}{2}\right) = -\frac{b^2t}{4}\left(\frac{2x_1}{b}\right) \tag{1}$$

For section 2 in the web

$$(A\bar{y})_2 = \frac{bt}{2}\left(-\frac{b}{2}\right) - ty\left(\frac{b}{2} - \frac{y}{2}\right)$$

or
$$(A\bar{y})_2 = -\frac{b^2t}{4}\left[1 + \left(\frac{2y}{b}\right) - \frac{1}{2}\left(\frac{2y}{b}\right)^2\right]\qquad(2)$$

For section 3 in the top flange

$$(A\bar{y})_3 = \frac{bt}{2}\left(-\frac{b}{2}\right) + tx_3\left(\frac{b}{2}\right)$$

or
$$(A\bar{y})_3 = -\frac{b^2t}{4}\left[1 - \left(\frac{2x_3}{b}\right)\right]\qquad(3)$$

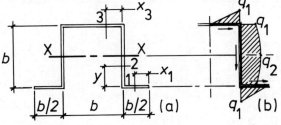

Figure 8.8

The shear flow distribution is obtained from equations 1 to 3 and shown (for half the cross-section) in figure 8.8b where

$$q_1 = -\frac{Q}{I_x} \times \frac{b^2t}{4}$$

and
$$q_2 = -\frac{Q}{I_x} \times \frac{3b^2t}{8}$$

The direction of the shear flow implied by the negative sign is shown in the figure. Note that in the other half of the cross-section, the flange shear flows will be reversed, since x_1 and x_3 will then be measured in the negative x-direction.

8.3 PROBLEMS FOR SOLUTION

1. A hollow beam is made by bending sheet steel 2 mm thick into a rectangular box of outside dimensions 100 mm × 80 mm. The joint is in the middle of a long side and is welded to complete the fabrication.

The beam is used as a short cantilever with the long side of the cross-section vertical. If a concentrated vertical load of 10 kN is placed at the free end, estimate the shear stress in the welded joint. (30·6 N mm^{-2})

2. Two wooden planks 1·5 m long and each 200 mm × 50 mm in cross-section are nailed together to form a tee-section beam which is used as a cantilever to carry a load of 5 kN at the free end. Determine the number of nails required if each nail is capable of resisting a shear force of 850 N.
(49)

3. The thin-walled section shown in figure 8.9 has a uniform thickness t. Show that the maximum shear stress produced by the shear force Q is given by

$$\frac{Q}{2\pi Rt} \left[1 + \frac{3\pi}{8} \right]$$

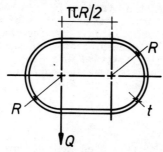

Figure 8.9

4. Figure 8.10 shows the cross-section of a beam. Determine the shear force which may be carried if the maximum shear stress is not to exceed 100 MN m^{-2}.
(1850 kN)

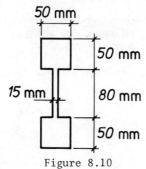

Figure 8.10

5. Determine the distribution of shear flow and the maximum shear stress in the singly symmetric I-section of figure 8.11 under a shear force of 100 kN acting through the centroid. Treat the section as thin.
(18·5 N mm^{-2})

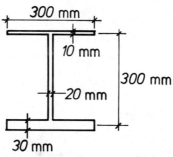

Figure 8.11

211

9 THICK CYLINDERS

In chapter 2 we examined the stresses in thin-walled cylinders subjected to internal pressure. The assumption of thin walls allowed
us to develop a simple analysis that ignored the variation of radial
stress. When the wall thickness becomes appreciable in proportion to
the cylinder radius, this variation in radial stress must be taken
into account in order to determine the correct distribution of
stresses. The theory is due to the French military engineer and
mathematician, G. Lamé.

9.1 LAMÉ'S THEORY

We shall assume that a thick-walled cylinder of internal radius a
and external radius b is subjected to an internal pressure p_i, the
external pressure being p_0. For the moment, longitudinal stresses
are assumed to be absent (this situation is achieved in the case of
a cylinder shrunk on to a shaft). We are thus dealing with a plane
stress problem.

Due to the internal pressure, tangential and radial stresses arise
in the cylinder wall. We start the analysis by determining the equilibrium equation for the stresses.

Figure 9.1a shows a cross-section through the cylinder.

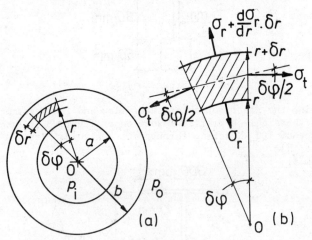

Figure 9.1

We consider the equilibrium of forces on the small element sub-
tending an angle $\delta\phi$ at the centre O and contained between the radii
r and $r + \delta r$ (figure 9.1b). An average value of the tangential
stress σ_t is assumed to act on the element. In fact σ_t is invariable
with respect to ϕ and varies with the radius, r. The radial stress

varies from σ_r on the inner face to $\sigma_r + (d\sigma_r/dr)\delta r$ on the outer face. The term $d\sigma_r/dr$ is the rate of change of radial stress with radius which is assumed to be linear over the small increment δr. As shown in figure 9.1b all stresses are taken to be tensile (positive).

We now resolve the forces on the element in the radial direction, then for equilibrium

$$\left(\sigma_r + \frac{d\sigma_r}{dr}\, \delta r\right)(r + \delta r)\delta\phi = \sigma_r r\, \delta\phi + 2\sigma_t\, \delta r \sin\frac{\delta\phi}{2}$$

After ignoring products of small quantities and noting that $\sin \delta\phi/2 \simeq \delta\phi/2$, we obtain

$$\sigma_t - \sigma_r - r\frac{d\sigma_r}{dr} = 0 \qquad\qquad (9.1)$$

The second part of the analysis requires the satisfaction of compatibility conditions. Figure 9.2 shows the radial displacements of the element in figure 9.1b.

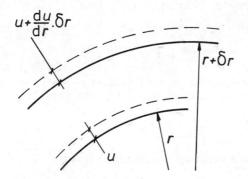

Figure 9.2

The inner face is displaced radially an amount u and the outer face an amount $u + du/dr\ \delta r$. The radial strain ε_r of the element is thus given by

$$\varepsilon_r = \frac{\delta r + \frac{du}{dr}\delta r - \delta r}{\delta r} = \frac{du}{dr} \qquad\qquad (a)$$

The tangential or circumferential strain ε_t at the radius r is given by

$$\varepsilon_t = \frac{\text{new circumference - old circumference}}{\text{old circumference}}$$

or $\quad \varepsilon_t = \dfrac{2\pi(r + u) - 2\pi r}{2\pi r} = \dfrac{u}{r} \qquad\qquad (b)$

The expressions for the strains are linked to the stresses by the elastic characteristics. Since σ_r and σ_t are principal stresses and

213

ε_r and ε_t are the corresponding principal strains, we have from equations 7.5 and 7.6

$$\sigma_r = \frac{E}{(1 - \nu^2)} \left(\frac{du}{dr} + \nu\frac{u}{r}\right) \tag{c}$$

and

$$\sigma_t = \frac{E}{(1 - \nu^2)} \left(\frac{u}{r} + \nu\frac{du}{dr}\right) \tag{d}$$

From equation c we have

$$\frac{d\sigma_r}{dr} = \frac{E}{(1 - \nu^2)} \left[\frac{d^2u}{dr^2} + \frac{\nu}{r^2} \left(r\frac{du}{dr} - u\right)\right]$$

Substituting for σ_r, σ_t and $d\sigma_r/dr$ in equation 9.1 we obtain

$$r\frac{d^2u}{dr^2} + \frac{du}{dr} - \frac{u}{r} = 0 \tag{9.2}$$

This is a straightforward second-order linear differential equation for which we try the solution $u = r^m$, then

$$rm(m - 1)r^{m-2} + mr^{m-1} - r^{m-1} = 0$$

or $\quad (m + 1)(m - 1)r^{m-2} = 0$

The indicial equation is satisfied if $m = +1$ or -1, thus the general solution of equation 9.2 takes the form

$$u = Ar + \frac{B}{r}$$

where A and B are constants of integration.

Substituting this result in equations c and d we obtain

$$\sigma_r = C - \frac{D}{r^2}$$

and

$$\sigma_t = C + \frac{D}{r^2} \tag{9.3}$$

where

$$C = \frac{AE}{1 - \nu} \quad \text{and} \quad D = \frac{BE}{1 + \nu}$$

Equations 9.3 are referred to as Lamé's equations for thick cylinders. C and D are constants for a particular cylinder under given pressures.

Example 9.1

A thick cylinder of internal diameter 30 mm and external diameter 50 mm is subjected to an internal pressure of 60 MN m^{-2} above the surrounding atmospheric pressure. Determine the distribution of tangential and radial stresses.

We use the given information to determine the constants C and D. From the first of equations 9.3 and noting that $\sigma_r \simeq -60$ MN m^{-2} when $r = 15$ mm and σ_r is approximately zero when $r = 25$ mm, we have

$$D = 625C \text{ and } D = 225(C + 60)$$

hence

$$C = 33 \cdot 75 \text{ MN m}^{-2} \text{ and } D = 21{,}094 \text{ N}$$

The radial and tangential stress distributions may now be drawn as shown in figure 9.3. The values of tangential stress at the inner and outer radii are $127 \cdot 5$ and $67 \cdot 5$ MN m^{-2} respectively.

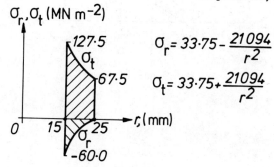

$$\sigma_r = 33 \cdot 75 - \frac{21094}{r^2}$$

$$\sigma_t = 33 \cdot 75 + \frac{21094}{r^2}$$

Figure 9.3

9.2 GRAPHICAL REPRESENTATION OF LAMÉ'S EQUATIONS - LAMÉ'S LINE

A convenient way of solving Lamé's equations is to plot the radial and tangential stresses against the reciprocal of the radius squared. Equations 9.3 are plotted in this way in figure 9.4a and b.

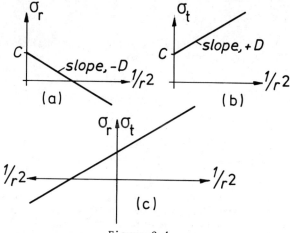

Figure 9.4

215

If the graph in figure 9.4a is rotated through 180° about the stress axis and combined with the graph in 9.4b we obtain the Lamé's line in 9.4c.

The previous example will now be solved again to illustrate the use of this method. It is usually referred to as the semi-graphical method since an exact plot of Lamé's line is not necessary, a sketch will suffice.

Figure 9.5 shows Lamé's line drawn for example 9.1.

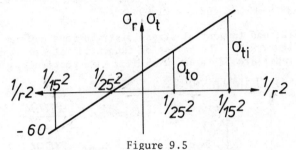

Figure 9.5

By similar triangles

$$\frac{60}{1/15^2 - 1/25^2} = \frac{\sigma_{t_o}}{2/25^2}$$

hence

$$\sigma_{t_o} = 60 \; \frac{(2)}{(25^2)} \; \frac{(15^2)(25^2)}{(25^2 - 15^2)} = 67 \cdot 5 \; \text{MN m}^{-2}$$

From the diagram

$$\sigma_{t_i} = \sigma_{t_o} + 60 = 127 \cdot 5 \; \text{MN m}^{-2}$$

9.3 STRAINS IN THICK CYLINDERS

The Lamé theory was developed as a two-dimensional or plane stress problem, the longitudinal strain at any radius being given by

$$\varepsilon_L = -\frac{\nu}{E}(\sigma_r + \sigma_t) \qquad \qquad \text{(a)}$$

thus from equations 9.3 we have

$$\varepsilon_L = -\frac{\nu}{E}\left(C - \frac{D}{r^2} + C + \frac{D}{r^2}\right) = -\frac{2\nu C}{E} \qquad \qquad \text{(b)}$$

The longitudinal strain is thus independent of the radius. This implies that plane sections of the cylinder remain plane under stress. It is therefore possible to apply a uniform longitudinal stress to the cylinder without affecting the distribution of the radial and tangential stresses. Thus Lamé's theory may be applied to cylinders

216

with closed ends. The uniform longitudinal stress condition does not hold close to the ends.

For a closed cylinder the longitudinal pressure force is distributed over the wall thickness, thus away from the ends, the longitudinal stress produced by a pressure p is

$$\sigma_L = \frac{p\pi a^2}{\pi(b^2 - a^2)} = p\frac{a^2}{(b^2 - a^2)} \tag{9.4}$$

the longitudinal strain is now given by

$$\varepsilon_L = \frac{1}{E}[\sigma_L - \nu(\sigma_r + \sigma_t)] \tag{c}$$

We are also interested in the diametral strain for thick cylinders. This, as we have seen previously (chapter 2), is the same as the tangential or circumferential strain (see also equation b of section 9.1 and figure 9.2). For a cylinder with closed ends, the tangential strain is given by

$$\varepsilon_t = \frac{1}{E}[\sigma_t - \nu(\sigma_r + \sigma_L)] \tag{d}$$

Example 9.2

A thick cylinder of length L, internal radius a and external radius b is closed by rigid end-plates. Determine the increase in storage volume when the contents are at a pressure p above atmospheric.

The initial internal volume V_0 is given by

$$V_0 = \pi a^2 L$$

If under pressure p the tangential strain at the inside surface is ε_{ti} and the longitudinal strain is ε_L the new internal volume is

$$V_0 + \Delta V_0 = \pi a^2 (1 + \varepsilon_{ti})^2 L(1 + \varepsilon_L)$$

Neglecting products of strains

$$V_0 + \Delta V_0 = \pi a^2 L(1 + 2\varepsilon_{ti} + \varepsilon_L)$$

or

$$\Delta V_0 = \pi a^2 L(2\varepsilon_{ti} + \varepsilon_L) \tag{1}$$

Now

$$\varepsilon_{ti} = \frac{1}{E}[\sigma_{ti} - \nu(-p + \sigma_L)]$$

and

$$\varepsilon_L = \frac{1}{E}[\sigma_L - \nu(-p + \sigma_{ti})]$$

By sketching the Lamé line we obtain the tangential stress at the inside surface as

$$\sigma_{ti} = p\left(\frac{b^2 + a^2}{b^2 - a^2}\right)$$

The longitudinal stress is given by

$$\sigma_L = p \frac{a^2}{(b^2 - a^2)}$$

Hence

$$\varepsilon_{t_i} = \frac{p}{E(b^2 - a^2)} \left[(b^2 + a^2) + \nu(b^2 - 2a^2) \right]$$

and $\quad \varepsilon_L = \frac{a^2 p}{E(b^2 - a^2)} (1 - 2\nu)$

Substituting in equation 1 above we have

$$\Delta v_0 = \frac{p\pi a^2 L}{E(b^2 - a^2)} \left[2b^2(1 + \nu) + 3a^2(1 - 2\nu) \right]$$

Example 9.3

The outside diameter of a thick steel cylinder is 200 mm. If the internal pressure is 50 MN m^{-2} what is the greatest value for the internal diameter if the maximum tangential stress is not to exceed 120 MN m^{-2}? Determine also the maximum shear stress and the increase in the internal diameter due to the pressure. E = 200 GN m^{-2} and ν = 0·3.

Figure 9.6a shows a cross-section through the cylinder. Figure 9.6b shows the Lamé line for the problem in terms of the diameters.

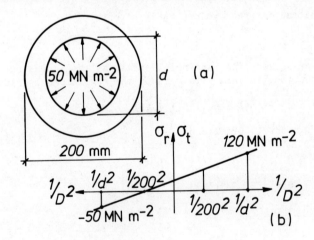

Figure 9.6

From the Lamé line

$$\frac{50}{1/d^2 - 1/200^2} = \frac{120}{1/d^2 + 1/200^2}$$

hence

218

$$d = 200\sqrt{\left(\frac{7}{17}\right)} = 128\cdot34 \text{ mm}$$

The uniform longitudinal stress σ_L away from the ends of the cylinder is given by

$$\sigma_L \frac{\pi}{4} (200^2 - d^2) = 50 \frac{\pi d^2}{4}$$

or $\quad \sigma_L = \dfrac{50d^2}{200^2 - d^2} = 35\cdot0 \text{ MN m}^{-2}$

The tangential strain at the inner surface is obtained from

$$\varepsilon_t = \frac{1}{E} (120 - \nu 35 + \nu 50)$$

or $\quad \varepsilon_t = \dfrac{1}{200 \times 10^3} (120 + 0\cdot3 \times 15) = 0\cdot62 \times 10^{-3}$

Since this is also the diametral strain at the inner surface we have

$$\Delta d = 0\cdot62d \times 10^{-3} = 0\cdot08 \text{ mm}$$

The maximum shear stress is at the inner surface and lies in the plane of the cross-section. By reference to chapter 7 we have

$$\tau_{max} = \frac{1}{2}[120 - (-50)] = 85 \text{ MN m}^{-2}$$

9.4 FORCE FITS

A common method of retaining a bush in a component is by using a force fit. The bush is made slightly larger in diameter than the hole into which it is to fit. On assembly, the bush is forced into the hole with a press. To determine the radial pressure on the bush (which will affect the press force required) and the stresses in the bush, Lamé's thick-cylinder theory is used. Note that the longitudinal stress is zero.

Example 9.4

A bronze bush having an external diameter of 70 mm and an internal diameter of 40 mm is press fitted into a recess in a body which is

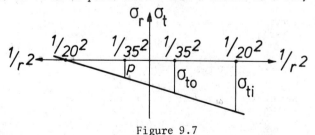

Figure 9.7

assumed to be perfectly rigid. If the diameter of the recess is 69·98 mm, find the radial pressure produced on the outside of the bush and the maximum tangential stress. Determine also the change in the internal diameter. E for bronze = 106 GN m^{-2} and ν = 0·35.

Figure 9.7 shows the Lamé line for the bush. Since a pressure p is applied at the outside surface and the pressure is zero at the inside, the tangential stresses are negative.

The tangential stress at the outer surface, σ_{t_0} is given by

$$\frac{\sigma_{t_0}}{1/20^2 + 1/35^2} = \frac{-p}{1/20^2 - 1/35^2}$$

or $\quad \sigma_{t_0} = -\left(\frac{35^2 + 20^2}{35^2 - 20^2}\right)p = -1\cdot97p$

The tangential strain at the outer surface is then

$$\varepsilon_t = \frac{1}{E}\left[-1\cdot97p - \nu(-p)\right]$$

Inserting values for E and ν we have

$$\varepsilon_t = -0\cdot0153p$$

if p is in GN m^{-2}. This is also the diametral strain at the outer surface which is given, thus

$$-0\cdot0153p = -\frac{0\cdot02}{70}$$

hence

$$p = \frac{0\cdot02}{70 \times 0\cdot0153} \text{ GN m}^{-2} = 18\cdot7 \text{ MN m}^{-2}$$

The maximum tangential stress in the bush occurs at the inner surface and from figure 9.7 is given by

$$\frac{\sigma_{t_i}}{2/20^2} = \frac{-p}{1/20^2 - 1/35^2}$$

or $\quad \sigma_{t_i} = -\frac{2 \times 35^2}{(35^2 - 20^2)}p = -55\cdot53 \text{ MN m}^{-2}$

Only one stress, σ_{t_i}, acts at the inner surface of the bush so that the corresponding tangential strain is given by

$$\varepsilon_{t_i} = \frac{\sigma_{t_i}}{E} = -\frac{55\cdot53}{106 \times 10^3} = -0\cdot524 \times 10^{-3}$$

the change in internal diameter, ΔD, is thus

$$\Delta D = -40 \times 0\cdot524 \times 10^{-3} = -0\cdot021 \text{ mm}$$

Example 9.5

A solid steel rod of diameter 64 mm is forced into a cylindrical bronze casing having an outside diameter of 100 mm. The resulting tangential stress on the outside of the casing is 35 MN m^{-2}. Determine

(a) the radial pressure between rod and casing

(b) the rise in temperature which would just eliminate the force fit.

For steel, E = 200 GN m^{-2}, ν = 0·28 and α = 12 × 10^{-6} deg K^{-1}.

For bronze, E = 110 GN m^{-2}, ν = 0·33 and α = 19 × 10^{-6} deg K^{-1}.

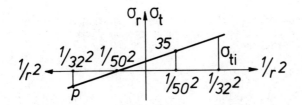

Figure 9.8

The Lamé line in figure 9.8 is drawn for the casing. The interface pressure p is therefore given by

$$\frac{p}{1/32^2 - 1/50^2} = \frac{35}{2/50^2}$$

or $\quad p = \dfrac{35(50^2 - 32^2)}{2 \times 32^2} = 25\cdot2$ MN m^{-2}

The tangential stress at the inside of the casing, σ_{ti} is obtained from figure 9.8, thus

$$\frac{\sigma_{ti}}{1/32^2 + 1/50^2} = \frac{35}{2/50^2}$$

or $\quad \sigma_{ti} = 35\dfrac{(50^2 + 32^2)}{2 \times 32^2} = 60\cdot2$ MN m^{-2}

To determine the stresses in the rod we refer to the Lamé equations, (9.3). Since the stresses cannot be infinite when r is zero, the constant D must be zero. Thus the radial and tangential stresses in the rod are independent of the radius and are equal in magnitude to the interface pressure p.

At the interface, the tangential strain in the casing is given by

$$\varepsilon_{tc} = \frac{1}{E_b}(\sigma_{ti} + \nu p)$$

or $\quad \varepsilon_{tc} = \dfrac{1}{110 \times 10^3}(60\cdot2 + 0\cdot33 \times 25\cdot2) = 0\cdot623 \times 10^{-3}$

221

The tangential strain in the rod is given by

$$\varepsilon_{t_r} = \frac{1}{E_s} \ (- \ p + \nu p)$$

or $\quad \varepsilon_{t_r} = \dfrac{- \ 25 \cdot 2}{200 \times 10^{-3}} \ (1 - 0 \cdot 28) = - \ 0 \cdot 091 \times 10^{-3}$

The change in the internal diameter of the casing is thus

$$\Delta D_c = + \ 64 \times 0 \cdot 623 \times 10^{-3} = + \ 0 \cdot 04 \ \text{mm}$$

and the change in the outside diameter of the rod is

$$\Delta D_r = - \ 64 \times 0 \cdot 091 \times 10^{-3} = - \ 0 \cdot 0058 \ \text{mm}$$

Let us now suppose that an increase in temperature of T K causes both rod and casing to expand to a common diameter D. For the free expansion of the rod

$$D = (64 + 0 \cdot 0058)(1 + 12T \times 10^{-6}) \ \text{mm}$$

similarly for the casing

$$D = (64 - 0 \cdot 04)(1 + 19T \times 10^{-6}) \ \text{mm}$$

Eliminating D between these two equations we obtain

$$T = 102 \cdot 2 \ \text{K}$$

9.5 COMPOUND CYLINDERS

The use of compound cylinders is a method of obtaining a more uniform distribution of stress under internal pressure and can be used for

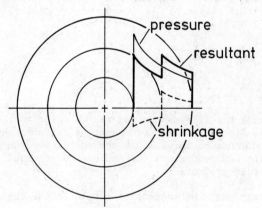

Figure 9.9

cylinders of comparatively short length. The method consists of shrinking one cylinder on to the outside of another. When the compound cylinder is subjected to internal pressure, the resultant stresses are obtained by the algebraic sum of those due to the internal pressure and those due to the shrinkage. This summing does not apply if parts of the cylinder have yielded. Figure 9.9 shows the separate and resultant tangential stresses in a compound cylinder.

9.5.1 Shrinkage Allowance

The stress distribution due to shrinkage is governed by the difference in the two cylinder diameters at the common interface. This difference in diameters is referred to as the shrinkage allowance.

Figure 9.10 shows a section through a compound cylinder. The common radial pressure at the interface is p. The tangential stress at the outer surface of the inner cylinder is σ_{to} while for the inner surface of the outer cylinder it is σ'_{ti}. Note that the prime applies to the outer cylinder. We shall suppose that the inner cylinder has elastic constants E and ν. For the outer cylinder the constants are E' and ν'.

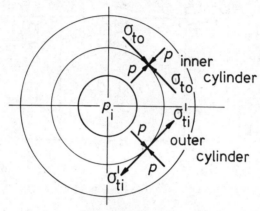

Figure 9.10

At the interface, the tangential strains are ε_t and ε_t' for the inner and outer cylinders respectively. Thus

$$\varepsilon_t = \frac{1}{E}[\sigma_{to} - \nu(p + \sigma_L)]$$

and $\quad \varepsilon'_t = \frac{1}{E'}[\sigma'_{t_i} - \nu'(p + \sigma_L)]$

where σ_L is the longitudinal stress which is assumed to be uniform and all stresses are taken as tensile.

If D is the interface diameter, the relative change in diameter at the interface or the shrinkage allowance S is thus given by

$$S = (\varepsilon'_t - \varepsilon_t)D$$

or $\quad S = \left[\frac{\sigma'_{ti}}{E'} - \frac{\sigma_{to}}{E}\right]D - \left[\frac{\nu'}{E'} - \frac{\nu}{E}\right](p + \sigma_L)D$

223

If both cylinders are of the same material, we have

$$S = (\sigma'_{t_i} - \sigma_{t_o}) \frac{D}{E}$$

Example 9.6

A steel cylinder of inside diameter 200 mm and 25 mm wall thickness is reinforced by shrinking on to it a second steel cylinder of 300 mm outside diameter. A pressure of 62 MN m^{-2} is applied to the inside of the compound cylinder. Calculate the shrinkage allowance so that, under pressure, the maximum tangential stresses in the inner and outer cylinders are the same and find the value of this stress. $E = 200$ GN m^{-2}.

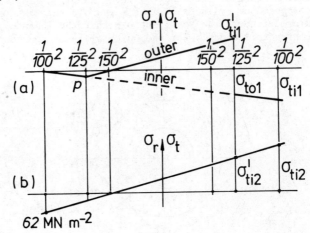

Figure 9.11

Figure 9.11a shows the Lamé lines for the two cylinders under shrinkage alone. Figure 9.11b is the common Lamé line for the additional stresses due to the internal pressure. The two cylinders may be treated as one from the point of view of internal pressure, since they are of the same material.

To meet the conditions of the problem we require that

$$\sigma'_{t_{i1}} + \sigma'_{t_{i2}} = \sigma_{t_{i1}} + \sigma_{t_{i2}} \tag{1}$$

where subscript 1 refers to shrinkage alone and subscript 2 refers to the internal pressure alone.

From figure 9.11a we have

$$\sigma'_{t_{i1}} = \frac{(150^2 + 125^2)}{(150^2 - 125^2)} p = 5 \cdot 54 p$$

where p is the radial pressure at the interface due to shrinkage alone; and

224

$$\sigma_{til} = - \frac{2 \times 125^2 p}{(125^2 - 100^2)} = -5 \cdot 55p$$

From figure 9.11b we obtain

$$\sigma'_{ti2} = 62\left(\frac{100}{125}\right)^2 \left(\frac{150^2 + 125^2}{150^2 - 100^2}\right) = 121 \cdot 0 \text{ MN m}^{-2}$$

and $\sigma_{ti2} = 62\left(\frac{150^2 + 100^2}{150^2 - 100^2}\right) = 161 \cdot 1 \text{ MN m}^{-2}$

Substituting in equation 1 we have

$$5 \cdot 54p + 121 \cdot 0 = - 5 \cdot 55p + 161 \cdot 1$$

hence $p = 3 \cdot 62$ MN m^{-2}.

The maximum tangential stress in the compound cylinder is thus

$$(\sigma_t)_{max} = 5 \cdot 54 \times 3 \cdot 62 + 121 \cdot 0 = 141 \cdot 0 \text{ MN m}^{-2}$$

The shrinkage allowance is given by

$$S = \frac{250}{200 \times 10^3} (\sigma'_{til} - \sigma_{tol}) \text{ mm}$$

From figure 9.11a

$$\sigma_{tol} = - \left(\frac{125^2 + 100^2}{125^2 - 100^2}\right)p = - 4 \cdot 55p$$

The shrinkage allowance thus becomes

$$S = \frac{250}{200 \times 10^3} (5 \cdot 54 + 4 \cdot 55) \ 3 \cdot 62 \text{ mm} = 0 \cdot 046 \text{ mm}$$

Figure 9.12 shows the separate and resultant stress distributions for the problem.

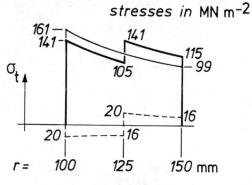

Figure 9.12

225

9.6 PROBLEMS FOR SOLUTION

1. A thick steel cylinder is required to hold fluid at a pressure of
18·5 MN m^{-2}. The outside diameter is to be 760 mm. Determine

    (a) the inside diameter so that the maximum shear stress shall not
        exceed 62 MN m^{-2}

    (b) the circumferential stresses at inner and outer surfaces

    (c) the increase in the internal diameter due to the pressure.

Take E = 205 GN m^{-2}, ν = 0·3. Neglect effect of longitudinal stress.
(636 mm, 105·5 and 86·8 MN m^{-2}, 0·353 mm)

2. Two steel cylinders are arranged concentrically one within the
other. They are of the same length and the ends are sealed off by
two concentric annular discs. The volume between the cylinders is
used to store fluid at gauge pressure p. The external diameters of
the cylinders are respectively 400 mm and 200 mm and the internal
diameters are 300 mm and 140 mm. The outside of the outer cylinder
and the inside of the inner cylinder are both exposed to atmospheric
pressure. Determine the value of p if the maximum shear stress in
either cylinder is not to exceed 100 MN m^{-2}.
(43·75 MN m^{-2})

3. A hollow cylinder having internal and external diameters of 89 mm
and 152·5 mm is subjected to an internal pressure of 20·7 MN m^{-2}.
The longitudinal strain was found to be 44 × 10^{-6}; this strain includes the effect of longitudinal stress. The same longitudinal
strain was obtained by subjecting the cylinder to an axial pull of
60 kN with no internal pressure. Use this information to determine
Poisson's ratio and Young's modulus for the cylinder material.
(0·267, 113 GN m^{-2})

4. A bronze bush having an external diameter of 300 mm is shrunk on
to a solid steel shaft of diameter 200 mm. If a temperature rise
of 100 K is just enough to eliminate the interference fit, determine
the original radial pressure at the interface.

    For steel, E = 208 GN m^{-2}, ν = 0·29 and α = 12 × 10^{-6} K^{-1}.

    For bronze, E = 112 GN m^{-2}, ν = 0·33 and α = 18 × 10^{-6} K^{-1}.
(20·2 MN m^{-2})

5. A thick cylinder is subjected to an internal pressure of 60 MN m^{-2}
which gives rise to longitudinal and tangential strains on the outside surface of 102·6 × 10^{-6} and 420·0 × 10^{-6} respectively. If the
external diameter is 50% greater than the internal diameter, use this
information to determine Poisson's ratio and Young's modulus for the
cylinder material.
(0·29, 195 GN m^{-2})

6. A compound pressure-vessel under an internal pressure of 80 MN m^{-2}
is made by shrinking a cylinder of outside diameter $4d$ and inside
diameter $2d$ on to another cylinder of the same material having an

outside diameter $2d$ and an inside diameter d. Determine the interface pressure due to shrinkage alone if the maximum shear stresses in
each cylinder are to be the same. Neglect the effect of longitudinal
stress.
(27 MN m^{-2})

7. A steel tube of length 200 mm and outside diameter 100 mm is
shrunk on to a long, solid steel shaft of diameter 50 mm. If the
limiting torque which may be transmitted between the tube and the
shaft is to be 20 kN m, determine the interface pressure between tube
and shaft and the shrinkage allowance required to produce this
pressure. The coefficient of friction between rod and shaft may be
taken as 0·2 and E = 200 GN m^{-2}.
(127·3 MN m^{-2}, 0·085 mm)

8. A compound cylinder consists of two steel cylinders, one being
shrunk on the other; the inside diameter is 100 mm, the outside diameter is 200 mm and the common diameter is 150 mm. If the internal
pressure in the compound cylinder is 110 MN m^{-2} find the radial
pressure at the interface which must be obtained by shrinkage if the
maximum tangential stresses in the inner and outer cylinders are to
be the same. Find also the shrinkage allowance if E = 200 GN m^{-2}.
(11·1 MN m^{-2}, 0·05 mm)

9. A steel cylinder 100 mm outside diameter and 75 mm inside diameter
is a good sliding fit at 15 K over a brass cylinder which is 75 mm
outside diameter and 50 mm inside diameter. If the compound cylinder
is now heated to 205 K, determine

   (a) the maximum stresses in brass and steel

   (b) the increase in diameter of the outer surface of the steel
       cylinder

   For steel, E = 200 GN m^{-2}, ν = 0·3 and α = 12 × 10^{-6} K^{-1}.

   For brass, E = 80 GN m^{-2}, ν = 0·33 and α = 21 × 10^{-6} K^{-1}.

(-129·0, 127·9 MN m^{-2}, 0·274 mm)

10. A thick cylinder is required to contain oil at a gauge pressure
of 60 MN m^{-2}. The external diameter is to be 250 mm. Determine the
necessary wall thickness if the maximum shear stress is not to
exceed 100 MN m^{-2}.

   What axial compressive force may now be applied to the pressur-
ised cylinder without violating the above limiting stress criterion?

   Neglect local stress effects at the ends of the cylinder and any
change in oil pressure due to the axial compression.
(46 mm, 2·95 MN)

10 COLUMNS

The analysis of column behaviour relies on the equation relating moment and curvature which we have already met in the study of beams (chapter 5). The general type of column carries both axial and transverse forces. The presence of the axial force and transverse deflexions produces bending moments in the column that are proportional to the transverse deflexions and although the column material may be linearly elastic, the resulting deflexions are non-linear with respect to the loads.

10.1 THE EULER COLUMN

During his investigations of the properties of the *elastica*, the Swiss mathematician, Leonhard Euler, was able to show (in 1744) that the least axial load to produce curvature in an initially straight strip of length L with pinned ends and no transverse loads is given by

$$P_E = \frac{\pi^2 EI}{L^2}$$

Euler used the exact curvature expression and showed further that if the axial load P is less than P_E, the strip remains straight and that as P is increased above P_E the transverse deflexion, Δ, of the strip also increases (see figure 10.2).

We shall now derive Euler's result for an initially straight pin-ended column. By pin-ended we mean that the ends are fixed in position but not in direction. The approximate expression for the curvature is used which is acceptable if slopes are small.

Figure 10.1 shows the column in equilibrium under the axial load P_E which is just sufficient to produce a small transverse deflexion.

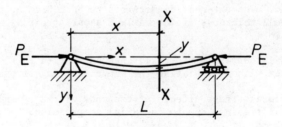

Figure 10.1

Taking the origin of the xy-coordinate system at the left-hand pin, the moment at section XX is given by

228

$$M_x = EI \frac{d^2y}{dx^2} = -P_E y$$

hence

$$\frac{d^2y}{dx^2} + \alpha_E^2 y = 0 \qquad\qquad\qquad (a)$$

where

$$\alpha_E = \sqrt{(P_E/EI)}$$

The general solution to the differential equation a is

$$y = A \cos \alpha_E x + B \sin \alpha_E x$$

where A and B are integration constants

Now $y = 0$ when $x = 0$ and L. From the first of these conditions we have $A = 0$ and the second gives

$$B \sin \alpha_E L = 0$$

B cannot be zero, since there would then be no transverse deflexion. We must therefore have that

$$\sin \alpha_E L = 0 \text{ or } \alpha_E L = \pi, \, 2\pi, \, 3\pi, \text{ etc.}$$

The first of these values of $\alpha_E L$ gives the first Euler critical load for the column

$$P_E = \frac{\pi^2 EI}{L^2} \qquad\qquad\qquad (10.1)$$

Under this axial load, the column deflexion is given by

$$y = B \sin \frac{\pi x}{L}$$

the higher valued solutions for $\alpha_E L$ give the deflected shapes

$$y = B \sin \frac{2\pi x}{L}, \; y = B \sin \frac{3\pi x}{L}, \text{ etc.}$$

and lead to critical loads $4P_E$, $9P_E$, etc.

The constant B is indeterminate and this implies that once the critical axial load P_E is reached, the column deflexions Δ cannot be determined precisely. If the approximate curvature expression continued to hold and if the material was infinitely elastic, the deflexion would be given by curve b of figure 10.2. In fact, these assumptions break down for quite small ratios of Δ/L.

The use of the exact curvature expression gives curve a of figure

10.2. The sudden branching from the straight to the curved form at the critical load P_E is referred to as a bifurcation (forking) of equilibrium.

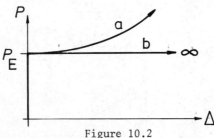

Figure 10.2

If equation 10.1 is divided by the cross-sectional area A of the column we obtain the critical Euler stress

$$\sigma_E = \frac{\pi^2 E}{(L/r)^2} \tag{10.2}$$

where $r = \sqrt{(I/A)}$ and L/r is referred to as the slenderness ratio of the column.

If the critical stress is plotted against the slenderness ratio, we obtain the Euler column curve in figure 10.3.

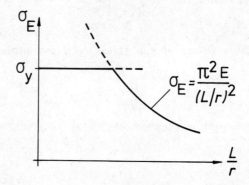

Figure 10.3

The critical stress expression (equation 10.2) is only true if the material remains elastic. A valid upper limit to the curve in figure 10.3 is therefore given by $\sigma_E \leq \sigma_y$.

10.1.1 The Importance of End Conditions

The end of a column may be fixed in position but not in direction (pinned), fixed in position and direction (built-in) or free (not fixed in position or direction). Various elastic restraints may be present that will result in intermediate end states.

Figures 10.4a, b and c show three axially loaded columns with different end conditions and their corresponding critical loads.

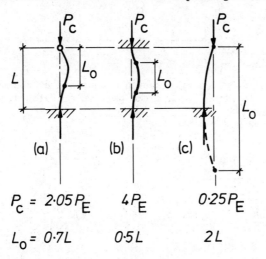

$$P_c = 2{\cdot}05\,P_E \qquad 4\,P_E \qquad 0{\cdot}25\,P_E$$

$$L_o = 0{\cdot}7L \qquad 0{\cdot}5L \qquad 2L$$

Figure 10.4

The basic method of analysis required to determine these critical loads is no different from that already used for the column
pinned at both ends. As an illustration, the result for the column
in figure 10.4a is derived in example 10.1.

Example 10.1

Determine an approximate value for the first critical load of an
initially straight, axially loaded column of length L which is
pinned at one end and built-in at the other.

Figure 10.5 shows the column with a small transverse deflexion
under the critical axial load. We note that since a moment is generated at the built-in end, a shear force Q is required at the
pinned end in order to hold it in position.

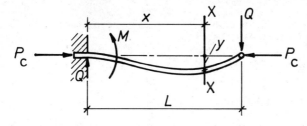

Figure 10.5

From the figure we have the moment at section XX as

$$M_x = EI\,\frac{d^2y}{dx^2} = -\,P_c y + Q(L - x)$$

231

or $\quad \dfrac{d^2y}{dx^2} + \alpha_c{}^2 y = \alpha_c{}^2 \dfrac{Q}{P_c}(L - x)$ $\hspace{3cm}$ (a)

where

$\qquad \alpha_c = \sqrt{(P_c/EI)}$

The general solution of equation a consists of the sum of the complementary function, y_1 and the particular integral y_2 where

$\qquad y_1 = A \cos \alpha_c x + B \sin \alpha_c x$

and $\quad y_2 = \dfrac{1}{(D^2 + \alpha_c{}^2)} \times \dfrac{\alpha_c{}^2 Q}{P_c} (L - x) = \dfrac{Q}{P_c} (L - x)$

hence

$\qquad y = A \cos \alpha_c x + B \sin \alpha_c x + \dfrac{Q}{P_c} (L - x)$

Now $y = 0$ at $x = 0$, thus

$\qquad A + \dfrac{QL}{P_c} = 0$ $\hspace{5cm}$ (b)

$y = 0$ at $x = L$, thus

$\qquad A + B \tan \alpha_c L = 0$ $\hspace{4.5cm}$ (c)

and $\quad \dfrac{dy}{dx} = 0$ at $x = 0$, thus

$\qquad \alpha_c B - \dfrac{Q}{P_c} = 0$ $\hspace{5cm}$ (d)

From equations b, c and d we obtain

$\qquad \alpha_c L = \tan \alpha_c L$ $\hspace{5cm}$ (e)

if B is not equal to zero.

The smallest value of $\alpha_c L$ (other than zero) that satisfies equation e is approximately 4·5, hence

$\qquad P_c \simeq (4{\cdot}5)^2 \dfrac{EI}{L^2} = 2{\cdot}05 \ P_E$

10.1.2 Effective Lengths

In column design, different end conditions are dealt with by reference to the effective length of the column, L_0. This is the length of pin-ended column that will fail at the same axial load as the column in question.

The effective length of a column is the length between points of

232

zero moment. Effective lengths are given for the columns in figure
10.4. Equations 10.1 and 10.2 are used to determine critical load
or stress with L_0 substituted for L. For the column of example 10.1

$$\frac{\pi^2 EI}{L_0^2} \simeq (4 \cdot 5)^2 \frac{EI}{L^2}$$

thus

$$L_0 \simeq \frac{\pi}{4 \cdot 5} L \simeq 0 \cdot 7L$$

10.2 REAL COLUMNS

The Euler critical load, P_E and the corresponding critical stress,
σ_E apply to an ideal column. They represent a theoretical upper
bound to column strength and have little significance in practical
column design.

Column behaviour is sensitive to imperfections and real columns
contain imperfections principally in the form of lack of initial
straightness, eccentricity of loading and locked-in stresses due to
manufacture. It is impossible to include the latter in a simple
analysis but their effect may be allowed for approximately by an
appropriate increase in the initial lack of straightness or eccentricity of loading.

10.2.1 The Initially Curved Column

Figure 10.6 shows an initially curved, pin-ended column of length L,
under axial load P. Under zero axial load, the column axis has a
deflected shape given by

$$y_0 = \Delta \sin \frac{\pi x}{L} \tag{a}$$

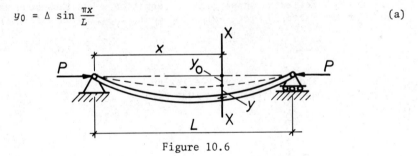

Figure 10.6

The presence of the axial load causes additional transverse deflexions y, thus the moment at section XX is given by

$$M_x = EI \frac{d^2 y}{dx^2} = -P(y + y_0)$$

or $$\frac{d^2 y}{dx^2} + \alpha^2 y = -\alpha^2 \Delta \sin \frac{\pi x}{L} \tag{b}$$

where $\alpha = \sqrt{(P/EI)}$.

233

The complementary function is

$$y_1 = A \cos \alpha x + B \sin \alpha x$$

and the particular integral is given by

$$y_2 = - \frac{\alpha^2 \Delta}{(D^2 + \alpha^2)} \sin \frac{\pi x}{L} = \frac{- \alpha^2 \Delta}{(- \pi^2/L^2 + \alpha^2)} \sin \frac{\pi x}{L}$$

The general solution of b is thus

$$y = A \cos \alpha x + B \sin \alpha x + \frac{P\Delta}{(P_E - P)} \sin \frac{\pi x}{L}$$

To determine the integration constants, we make use of the boundary conditions $y = 0$ at $x = 0$ and L. The first of these gives $A = 0$ and the second gives

$$B \sin \alpha L = 0$$

If $\sin \alpha L = 0$ we obtain P equal to the first Euler critical load P_E which cannot be the correct solution for this problem since we are dealing with an imperfect column. Hence we must have $B = 0$, thus

$$y = \frac{P\Delta}{P_E - P} \sin \frac{\pi x}{L}$$

and the total column deflexion is

$$y + y_0 = \frac{P_E \Delta}{P_E - P} \sin \frac{\pi x}{L}$$

The initial deflected shape given by equation a has thus been magnified by the factor $P_E/(P_E - P)$. The central deflexion Y is given by

$$Y = \frac{P_E}{P_E - P} \Delta \qquad\qquad\qquad (c)$$

Figure 10.7 shows the central deflexion for the column plotted against the axial load for different values of Δ.

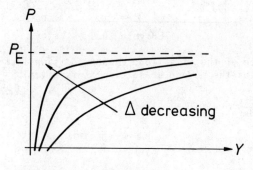

Figure 10.7

234

The deflexion has a definite value for each value of axial load and there is no bifurcation of equilibrium. For finite values of Δ, the first Euler critical load P_E is reached for infinite values of Y. In practice, of course, material breakdown would occur long before this point. When Δ approaches zero we obtain the result for the ideal column.

The failure of an imperfect column is determined by strength considerations and not by the attainment of a critical load. We assume that the column is on the point of failure when the cross-section ceases to be fully elastic. Thus the maximum stress in the column is not to exceed the yield stress of the material.

The maximum stress occurs at the mid-point and is composed of the direct compressive stress and the maximum compressive bending stress, thus at failure

$$\frac{P}{A} + \frac{PY}{Z} = \sigma_Y$$

where Z is the least value of the elastic section modulus, A is the cross-sectional area and σ_Y is the yield stress.

After substituting for Y we have

$$(\sigma_Y - \sigma)(\sigma_E - \sigma) = \sigma\sigma_E \left(\frac{A\Delta}{Z}\right) \tag{d}$$

where $\sigma = P/A$.

The solution of equation d for the failure stress σ is the lesser root, thus

$$\sigma = \left[\frac{\sigma_Y + \sigma_E(1 + \eta)}{2}\right] - \sqrt{\left\{\left[\frac{\sigma_Y + \sigma_E(1 + \eta)}{2}\right]^2 - \sigma_Y\sigma_E\right\}} \tag{10.3}$$

where $\eta = A\Delta/Z$.

Equation 10.3 is referred to as the Perry curvature formula after one of the men who derived it in 1886 (the other was Ayrton). Experi-

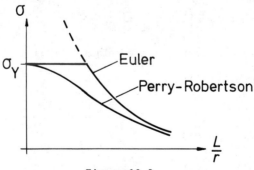

Figure 10.8

235

mental work by Robertson in 1925 gave a value to η and the formula now has Robertson's name associated with it.

The Perry-Robertson formula has been used in British structural steel design codes for many years.*

Figure 10.8 shows the column design curve drawn from equation 10.3 compared with the Euler curve.

10.2.2 The Eccentrically Loaded Column

Figure 10.9 shows an initially straight pin-ended column of length L which carries a load P whose line of action is parallel to and a distance e from the column axis.

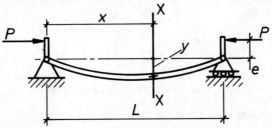

Figure 10.9

If the transverse deflexion from the original position of the axis is y, the moment at section XX is

$$M_x = EI \frac{d^2y}{dx^2} = - P(y + e)$$

or $\quad \dfrac{d^2y}{dx^2} + \alpha^2 y = - \alpha^2 e$

the general solution of which is

$$y = A \cos \alpha x + B \sin \alpha x - e \qquad \text{(a)}$$

From the boundary conditions $y = 0$ at $x = 0$ and L, we have

$$A = e \text{ and } B = e\left(\frac{1 - \cos \alpha L}{\sin \alpha L}\right)$$

The maximum deflexion in the column is thus

$$Y = e\left(\sec \frac{\alpha L}{2} - 1\right)$$

---

*In BS 153 $\quad$ η $= 0 \cdot 3L/100r$

In BS 449 $\quad$ η $= 0 \cdot 3 \ (L/100r)^2$

236

As with the initially curved column, we assume that failure is associated with the onset of yield. If σ is the failure stress we have

$$\sigma_Y = \sigma + \frac{P}{Z}\left[e + e\left(\sec \frac{\alpha L}{2} - 1\right)\right]$$

or $\sigma_Y = \sigma\left[1 + \left(\frac{Ae}{Z}\right) \sec \frac{\pi}{2}\sqrt{\left(\frac{\sigma}{\sigma_E}\right)}\right]$ (10.4)

It is difficult to obtain an explicit expression for σ from equation 10.4 and hence the eccentricity assumption is little used in column design. The relationship between end load and central deflexion takes a similar form to that shown in figure 10.7.

Example 10.2

A column of length 6 m has a least radius of gyration of 10 cm, a cross-sectional area of 800 cm^2 and an elastic section modulus of 4800 cm^3. Compare the axial loads to produce yielding in the column if

(a) the initial centreline of the column is a half sine wave with a maximum bow of 6 mm

(b) the column is initially straight but the axial load is applied with an eccentricity of 6 mm at each end.

Assume the yield stress is 250 MN m^{-2} and E = 200 GN m^{-2}.

The column has a slenderness ratio given by

$$\frac{L}{r} = \frac{600}{10} = 60$$

hence the first Euler critical stress is

$$\sigma_E = \frac{\pi^2 E}{(60)^2} = 548 \cdot 3 \text{ MN } m^{-2}$$

(a) The value of η in equation 10.3 is

$$\eta = \frac{800 \times 0 \cdot 6}{4800} = 0 \cdot 1$$

From equation 10.3 we find that the limiting stress for a is 214·7 MN m^{-2}. The corresponding axial load is therefore 17·18 MN.

(b) The value of Ae/Z for the eccentrically loaded column is also 0·1. After substituting in equation 10.4 we obtain

$$\sigma\left[1 + 0 \cdot 1 \times \sec \frac{\pi}{2}\sqrt{\left(\frac{\sigma}{\sigma_E}\right)}\right] = \sigma_Y$$

This equation may be solved for σ by making use of the following approximate relationship

$$\sec \theta = \frac{1 + 0\cdot26\left(\frac{2\theta}{\pi}\right)^2}{1 - \left(\frac{2\theta}{\pi}\right)^2} \quad \text{for } 0<\theta<\frac{\pi}{2}$$

hence

$$\sec \frac{\pi}{2}\sqrt{\left(\frac{\sigma}{\sigma_E}\right)} = \frac{\sigma_E + 0\cdot26\sigma}{\sigma_E - \sigma}$$

thus

$$\sigma = \frac{(\sigma_Y + 1\cdot1\sigma_E)}{1\cdot948} - \sqrt{\left[\left(\frac{\sigma_Y + 1\cdot1\sigma_E}{1\cdot948}\right)^2 - 1\cdot0267\ \sigma_E\sigma_Y\right]}$$

since $\sigma_Y = 250$ MN m^{-2} and $\sigma_E = 548\cdot3$ MN m^{-2} we have $\sigma = 211\cdot9$ MN m^{-2}. The corresponding end thrust is thus $16\cdot95$ MN. The assumption of an eccentricity in the application of the load is therefore slightly more severe for the given value of the slenderness ratio than the use of an initial sinusoidal curvature.

10.2.3 Column Design Formulae

A very large number of different column design formulae exist, most of which have an empirical or an intuitive background. The most relevant formula in current use is the Perry-Robertson formula which as we have seen in section 10.2.1 has a sound rational basis.

It is only recently that a thorough analysis of the column problem has been achieved, full allowance having been made for geometrical imperfections and locked-in stresses. Calculations carried out using a computer have resulted in a set of multiple column curves, each of which apply to a particular class of column cross-section.

10.3 BEAM-COLUMNS

As the name implies, a beam-column is a structural member which carries both transverse loads and thrust. The technique of solution is no different from that employed for the simple column; we determine the general expression for the moment and solve the resulting second-order differential equation for the deflexion. If the bending moment in the beam-column is required, it is also possible to determine a differential equation for moment. Two typical beam-column examples will now be treated.

Example 10.3

A pin-ended beam-column of length L and flexural rigidity EI under an axial thrust P is subjected to a uniformly distributed transverse load of w per unit length. Determine the maximum bending moment and the maximum deflexion.

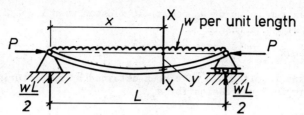

Figure 10.10

Figure 10.10 shows the loaded column. The moment at XX is

$$M_x = EI \frac{d^2y}{dx^2} = -Py - \frac{wL}{2}x + \frac{wx^2}{2}$$

hence

$$\frac{d^2y}{dx^2} + \alpha^2 y = \frac{-w}{2EI} x(L - x)$$

The general solution is

$$y = A \cos \alpha x + B \sin \alpha x - \frac{wx}{2\alpha^2 EI}(L - x) - \frac{w}{\alpha^4 EI}$$

Now $y = 0$ at $x = 0$ and L, thus

$$A = \frac{w}{\alpha^2 P} \quad \text{and} \quad B = \frac{w}{\alpha^2 P} \tan \frac{\alpha L}{2}$$

The maximum deflexion, Y occurs at $x = \frac{L}{2}$, thus

$$Y = \frac{w}{\alpha^2 P} \sec \frac{\alpha L}{2} - 1 - \frac{wL^2}{8P}$$

To determine the maximum bending moment, we return to the expression for M_x. If this is differentiated twice with respect to x, we have

$$\frac{d^2 M_x}{dx^2} = -P \frac{d^2 y}{dx^2} + w$$

but $\frac{d^2y}{dx^2} = \frac{M_x}{EI}$

thus $\frac{d^2 M_x}{dx^2} + \alpha^2 M_x = w$

hence

$$M_x = A \cos \alpha x + B \sin \alpha x + \frac{w}{\alpha^2}$$

Since $M_x = 0$ at $x = 0$ and L, we have

$$A = \frac{-w}{\alpha^2} \quad \text{and} \quad B = \frac{-w}{\alpha^2} \tan \frac{\alpha L}{2}$$

239

The maximum value of M_x occurs at midspan when $x = L/2$, thus

$$(M_x)_{max} = \frac{w}{\alpha^2}\left(1 - \sec\frac{\alpha L}{2}\right)$$

The same result could have been obtained by substituting Y for y in the original expression for M_x.

Example 10.4

A pin-ended beam-column of length L is subjected to an axial load P together with sagging moments M and βM at the ends, $(\beta < 1)$. If both applied moments tend to increase the curvature of the column, obtain a general expression for the deflexion and the magnitude and position of the maximum moment.

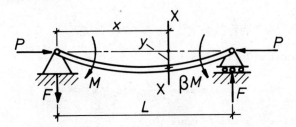

Figure 10.11

Figure 10.11 shows the loaded beam-column. Since the end moments are unequal, vertical reactions will be generated at the supports. The moment at section XX is

$$M_x = EI\frac{d^2y}{dx^2} = -Py - M + Fx$$

or $\quad EI\frac{d^2y}{dx^2} = -Py - M\left[1 - (1 - \beta)\frac{x}{L}\right]$

since

$$F = (1 - \beta)\frac{M}{L}$$

The differential equation for the deflexion is thus

$$\frac{d^2y}{dx^2} + \alpha^2 y = \frac{-M}{EI}\left[1 - (1 - \beta)\frac{x}{L}\right]$$

and the general solution is

$$y = A\cos\alpha x + B\sin\alpha x - \frac{M}{\alpha^2 EI}\left[1 - (1 - \beta)\frac{x}{L}\right]$$

From the boundary conditions, $y = 0$ at $x = 0$ and L we obtain

240

$$A = \frac{M}{\alpha^2 EI} = \frac{M}{P}$$

and $\quad B = \frac{M}{P}\left(\frac{\beta - \cos \alpha L}{\sin \alpha L}\right)$

thus $\quad y = \frac{M}{P}\left[\frac{\sin \alpha(L - x) + \beta \sin \alpha x}{\sin \alpha L} + (1 - \beta)\frac{x}{L} - 1\right]$

To determine the bending moment we differentiate the expression for M_x twice with respect to x, then

$$\frac{d^2 M_x}{dx^2} + \alpha^2 M_x = 0$$

hence

$$M_x = A \cos \alpha x + B \sin \alpha x$$

but $M_x = M$ when $x = 0$ and $M_x = \beta M$ when $x = L$, thus

$$A = M \text{ and } B = M\left(\frac{\beta - \cos \alpha L}{\sin \alpha L}\right)$$

hence

$$M_x = M\left[\cos \alpha x + \left(\frac{\beta - \cos \alpha L}{\sin \alpha L}\right)\sin \alpha x\right]$$

The position of the maximum bending moment in the beam-column will occur at the point where the shear force is zero. Let this point be x_m from the left-hand end then

$$\left(\frac{dM_x}{dx}\right)_{x = x_m} = M\left[- \alpha \sin \alpha x_m + \left(\frac{\beta - \cos \alpha L}{\sin \alpha L}\right)\alpha \cos \alpha x_m\right] = 0$$

or

$$x_m = \frac{1}{\alpha} \tan^{-1}\left(\frac{\beta - \cos \alpha L}{\sin \alpha L}\right)$$

Substituting x_m for x in the final expression for M_x, and noting that $(\beta - \cos \alpha L/\sin \alpha L) = \tan \alpha x_m$ we obtain the maximum moment as

$$(M_x)_{max} = M \sec \alpha x_m$$

10.4 PROBLEMS FOR SOLUTION

1. An initially straight column of length L and uniform flexural rigidity EI is built-in at the base and in the unloaded state is inclined to the vertical at a small angle θ.

241

If a vertical load P is applied to the free end, prove that the maximum moment in the column is

$$\theta \sqrt{(PEI)} \times \tan L \sqrt{\left(\frac{P}{EI}\right)}$$

2. A column AB of length L carries an axial thrust P. End A is pinned in position in such a way as to produce no rotational restraint. End B is also pinned in position but offers a rotational restraint $k\theta$, where θ is the slope at B.

Show that the critical load satisfies the expression

$$PL = k(\alpha L \cot \alpha L - 1)$$

where $\alpha = \sqrt{(P/EI)}$.

3. A beam-column under axial load P has a sagging moment M applied at one end and a hogging moment $M/2$ at the other. If the axial load $P = 0 \cdot 6P_E$, find the position and magnitude of the maximum moment. P_E is the first Euler critical load for the column under axial thrust alone. $(0 \cdot 155L, 1 \cdot 075M)$

4. An initially straight uniform elastic column of length L and appropriate second moment of area I is acted upon by axial loads P together with a moment M at one end only. Show that the maximum bending moment at any point in the column will not exceed M if

$$P \le \frac{\pi^2 EI}{4L^2}$$

Assume that the ends of the column are fixed in position but not in direction.

5. A beam-column of length L carries an end-thrust P and a moment M at one end only.

Determine the maximum moment in the column if (a) $P = 0 \cdot 2P_E$ and (b) $P = 0 \cdot 5P_E$ where P_E is the first Euler critical load for the column under axial load alone.

For what value of P would the maximum column moment appear at $L/4$ from the end carrying the moment M?
$(M, 1 \cdot 26 M, 4P_E/9)$

6. A beam-column of length L under axial load P is built-in at the ends. Show that when a concentrated lateral load W is applied at midspan the greatest moment in the column is reached simultaneously at the ends and at midspan. Show further that the magnitude of this moment is

$$\frac{W}{2\alpha} \tan \frac{\alpha L}{4} \text{ where } \alpha = \sqrt{(P/EI)}$$

7. A long thin column of length L is initially straight. It is then

loaded by an end thrust P acting with eccentricity e. The resulting deflexion at the centre is partially resisted by a spring which exerts a reaction $c\Delta$ where c is a constant and Δ is the deflexion of the column from the straight.

Show that

$$\Delta = \frac{2\alpha Pe\,(\sec \frac{\alpha L}{2} - 1)}{2\alpha P - c\left[\frac{\alpha L}{2} - \tan \frac{\alpha L}{2}\right]}$$

where $\alpha^2 = P/EI$

8. An initially straight column of length L carries an axial load P. There is a lateral distributed load which varies uniformly from w_0/unit length at one end to zero at the other. Show that the position of the maximum moment is given by

$$\cos \alpha x_m = \frac{\sin \alpha L}{\alpha L}$$

where x is measured from the end carrying zero distributed load and $\alpha^2 = P/EI$.

If P is 81% of the first Euler critical load, P_E, find the position and magnitude of the maximum moment.
$(0\cdot517L, \; -0\cdot34\; w_0L^2)$

9. An initially straight, slender column of uniform section and length L has pinned ends through which it is axially loaded by a force P, $(P_E < P < 4P_E)$. Equal moments M are applied at the ends to maintain equilibrium, the slope at each end then being ϕ radians.

Show that

(a) $M = \dfrac{P\phi}{\alpha} \cot \dfrac{\alpha L}{2}$

(b) the maximum deflexion

$\Delta = \dfrac{\phi}{\alpha} \tan \dfrac{\alpha L}{4}$

where $\alpha = \sqrt{(P/EI)}$

APPENDIX

PROPERTIES OF AREAS BOUNDED BY PARABOLAE

The curve $y = ax^2$ shown in figure A.1 divides the rectangle ORST into the areas A_1 and A_2.

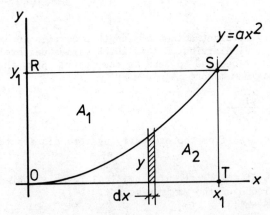

Figure A.1

The area A_2 between the curve and the x-axis is given by

$$A_2 = \int_0^{x_1} y \, dx = \int_0^{x_1} ax^2 \, dx = ax_1^3/3$$

The total area of the rectangle ORST is

$$A_1 + A_2 = y_1x_1 = ax_1^3$$

thus $A_1 = \dfrac{2}{3} ax_1^3$

or $A_1 = \dfrac{2}{3} (A_1+A_2)$ and $A_2 = \dfrac{1}{3} (A_1+A_2)$

Let the centroid of the area A_2 be at a distance $\bar{x}_2$ from the origin O, then

$$A_2\bar{x}_2 = \int_0^{x_1} xy \, dx = \int_0^{x_1} ax^3 \, dx = ax_1^4/4$$

But $A_2 = ax_1^3/3$

thus $\bar{x}_2 = 3x_1/4$

Now the centroid of the total area (A_1+A_2) is at a distance

244

$x_1/2$ from the origin. Thus if the centroid of the area A_1 is at a distance $\bar{x}_1$ from the origin we have

$$(A_1+A_2) \frac{x_1}{2} = A_1\bar{x}_1 + A_2\bar{x}_2$$

or $\quad (ax_1{}^3) \frac{x_1}{2} = \left(\frac{2}{3} ax_1{}^3\right) \bar{x}_1 + \left(\frac{1}{3} ax_1{}^3\right) \frac{3x_1}{4}$

hence

$$\bar{x}_1 = \frac{3x_1}{8}$$

FURTHER READING

S. P. Timoshenko, *History of Strength of Materials*, McGraw-Hill, New York, 1953

S. P. Timoshenko and J. M. Gere, *Mechanics of Materials*, Van Nostrand Reinhold, New York, 1972

J. Case and A. H. Chilver, *Strength of Materials and Structures*, Edward Arnold, London, 1971

J. E. Gordon, *The New Science of Strong Materials*, Penguin, Harmondsworth, 1968

W. D. Biggs, *The Mechanical Behaviour of Engineering Materials*, Pergamon, Oxford, 1965

INDEX